D1400479

Analytical and Computer Cartography

Analytical and Computer Cartography

Keith C. Clarke

Hunter College—City University of New York

 Prentice Hall, Englewood Cliffs, New Jersey 07632

Clarke, Keith C., 1955-
    Analytical and computer cartography / Keith C. Clarke.
        p.   cm.
    Includes bibliographical references.
    ISBN 0-13-033481-2
    1. Cartography--Data processing.   I. Title.
    GA102.4.E4C377 1990   89-71104
    526'.0285--dc20   CIP

Editorial/production supervision: Edward Thomas
Cover design: Ben Santora
Manufacturing buyer: Paula Massenaro

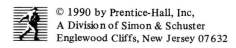 © 1990 by Prentice-Hall, Inc,
A Division of Simon & Schuster
Englewood Cliffs, New Jersey 07632

Printed in the United States of America

10 9 8 7 6 5 4 3 2

ISBN 0-13-033481-2

Prentice-Hall International (UK) Limited, *London*
Prentice-Hall of Australia Pty. Limited, *Sydney*
Prentice-Hall Canada Inc., *Toronto*
Prentice-Hall Hispanoamericana, S.A., *Mexico*
Prentice-Hall of India Private Limited, *New Delhi*
Prentice-Hall of Japan, Inc., *Tokyo*
Prentice-Hall of Southeast Asia Pte. Ltd., *Singapore*
Editora Prentice-Hall do Brasil, Ltda., *Rio de Janeiro*

For those I have taught,
for those I will teach,
and to those who taught me.

Contents

Foreword

In the present century, cartographers have seen the adoption of the metric system, designation of Greenwich as the prime meridian, specification of an international sequence of colors for terrain maps, the beginning of an international millionth map of the world, the almost universal adoption of plane coordinate systems for local surveys ... and the list goes on. But one is not allowed to write such long sentences. Still it is necessary to mention automobiles and road maps, airplanes and aeronautical charts, and aerial photography and photogrammetry. These latter items have been extended to satellite altitudes and to wavelengths—radar and infrared—beyond human sensing capabilities. The electronic revolution has led to the replacement of triangulation by trilateration, and of course, to computers.

The statistical information collection agencies have kept up, providing more data for thematic maps and atlases, until recently scribed on modern stable plastics, but increasingly generated by computer graphics software. Getting the world into the computer is still a problem, but more and larger data bases are fast becoming available. For those who know how to use it there has never been available nearly as much knowledge about geography. This is true at the local, regional, and international level. The rapidly growing geographic information systems contain formalizations of cartographic concepts which have evolved over thousands of years. The topological structuring of geographic data is so basic that it almost had to be rediscovered. With more and more of the solutions to everyday geographical problems solvable by algorithmic procedures—minimal path routines combined with a digital street network stored in an automobile provide one currently implementable example—those tasks which humans perform more quickly still rely on the impressive processing capabilities of the visual cortex. Thus modern cartography requires both the artistic for geographic illustration and the analytic for solving geographic problems. The introductory period of computer assisted

cartography dealt with mathematically simple, but tedious, operations—survey adjustments, ellipsoidal calculations, map projections, replacement of manual drafting by
plotters. More difficult tasks are now being approached—for example map name placement or map generalization—using systems of expert rules which attempt to mimic the
intuitive heuristics used by well trained cartographers. All of this, and more, is the
domain of analytical and computer cartography.

Waldo R. Tobler
Santa Barbara

Preface

I wrote this book because for five years I have been teaching a class entitled *Analytical and Computer Cartography* waiting for a textbook I could use, hoping that somebody else would write one. Since I arrived in New York in 1982, I have been tempted to start work on a text several times. I would like to thank Dr. Alan Strahler, now of Boston University, for talking me out of starting this book at the wrong time. I believe that now is the right time.

Waldo Tobler stated that his lecture notes for *Analytical Cartography* at the University of Michigan had a half-life of five years. This, clearly, was a conservative estimate. During my years at Hunter College, the set of lecture notes which evolved into this book was overhauled several times, and has been reworked yet again in the preparation of what I present here. This would have been impossible without the many students who have taken my classes, have provided insightful questions, have been quick to find the holes in my arguments, and who, more than any others, have encouraged me to produce this book.

Many individuals contributed to this book. I would like to thank Mark Bosworth and Karen Mays for transcribing my lectures, Joy Chen for assisting in research, David Lamb of Queens University for some valuable advice on troff and macros, and the Academic Computing Services of Hunter College, for lending me the Apple Macintosh on which I prepared most of the diagrams. The following colleagues provided valuable reviews; Marc Armstrong, Tim Nyerges, Tony Williams, Barbara Buttenfield, Ryan Rudnicki, J. Ronald Eastman, Jon Kimerling, Nina Lam, and Nick Chrisman. Of these, I would particularly like to thank Nick Chrisman, whose enthusiasm and depth of experience made his reviews almost artworks in themselves. At Prentice-Hall, Dan Joraanstad and Ed Thomas were important contributors to the organization and style of this book. Finally, thanks are due to Waldo Tobler, who provided both the foreword and the inspiration for analytical cartography.

I would also like to thank the students, both graduate and undergraduate, who have suffered through my programming class, *Computer Programming for Geographic Applications*. During the teaching of this class, I have reinforced my belief that the cartographer who does not command his or her tools becomes a slave to those tools. The fact is that too few cartographers, and far too few geographers, program. Even fewer can program in the structured languages which have swept the programming world, and fewer still are aware of the commonsense breakthroughs in software engineering which have made programming a more effective and efficient process than it was even a few years ago. In keeping with this philosophy, the programs in this book use a common set of data structures (presented in Chapter 5), are in the highly portable C programming language, and use the graphics routines of the graphical kernel system (GKS), the international graphics programming standard.

Cartography should go beyond training individuals to use the tools of the computer age, and should teach students how to build better mapping systems. To achieve this, students need a firm foundation of cartographic principles and a high set of aesthetic standards, which in the past have been taught only by manual cartographers. Over this foundation should come a mastery of the tools of computer cartography and an understanding of the power of analytical cartography. One barrier to this has been that formed by the traditional academic disciplinary boundaries. An active field of new study has a healthy ability to form its own links among tools, methods, and theory. Part of my motive in writing this book is to present such a link between cartography and computer science. The greater part of my motive, however, is to encourage cartographers to make better maps.

K.C.C.
New York City

1

The Computer and Cartography

1.1 THE COMPUTER AT THE CENTER OF CARTOGRAPHY

Cartography is a discipline as old as mankind and as young as today's newspaper. Why is this? Cartography is old because as a means of expression, the map probably predates many other forms of human communication, and maps survive which are several thousand years old. Cartography is young because it is a discipline which has been subjected to a series of revolutions in innovative technology. At first, these revolutions came separated by centuries, and in the case of the era following the decline of Greek and Roman cartography most of the cumulative knowledge of mapmaking was forgotten. Since the advent of the digital computer, however, we have become used to cartography as being in a state of almost constant technological revolution.

Why has this technological tool surpassed all others in the history of the discipline? How have cartographers adapted their discipline to this constant state of flux? And what will cartography look like in the years of innovation to come? In this book, we shall examine the computer revolution in cartography from the standpoint of how cartography has changed, and more important, from the standpoint of how it has remained the same. As we shall see, the revolution has shaped a new cartography, in which the specifics of technology appear to be overwhelming.

Fortunately, the revolution has also sent cartography back to its historical and mathematical roots, and therefore has made the basic principles of cartography as much in demand, if not more so, as at any previous time. When we think of the technology of mapmaking, the center of activity now lies within computer cartography. In the past, we thought of cartographers as scribes with quill pens scratching out maps of the world. This is simply no longer the way it is done. Mapmaking technologies have come and gone and the simple origins are now very distant. Cartographers have changed the way they see the creation of maps and the role of mapmaking itself within cartography. This has been the case in both manual and computer cartography, for cartography is a set of skills and a body of theory, and the theory remains the same independent of what particular technology one happens to use to make any particular map.

This is why this book has the title *Analytical and Computer Cartography*, for there are two interlinked themes. The first theme, *analytical cartography* (Tobler, 1976), deals with the theoretical and mathematical background behind cartography and the rules cartographers employ in the mapping process. The other theme is hands-on *computer cartography*, the particular set of methods and techniques which the current technology uses to produce maps. If we use the term *computer cartography,* then in the past we have had quill-pen cartography, mapping pen cartography, scribing cartography, and photogrammetric cartography, for these are some of the previous technologies we have used in the science of mapmaking. This book will instruct the reader in mapmaking using the most current tool, the digital computer.

This does not necessarily affect the body of theory behind cartography, but rather emphasizes and reemphasizes the cartographic lessons of the past. For example, for most map projections, the underlying equations and transformational geometry were worked out, sometimes to perfection, in previous cartographic eras. Today the equations still work whether we produce the map by hand construction or by computer graphics. A typical laboratory for computer cartography may contain microcomputers with graphics cards, high-resolution color monitors, digitizers, color plotters and printers rather than drafting tables and sinks. While a student in a class on computer cartography may use such a laboratory now, we probably (in fact definitely) will have something different just a few years from now. This is the nature of scientific and technological revolutions. However, most of the theory and principles presented and reviewed in this book will work just as well using the technology of next year or the next century, just as most of the principles worked when they were derived centuries ago.

1.2 STAGES OF ADAPTATION OF THE COMPUTER INTO CARTOGRAPHY

Morrison (1980) stated that there are three stages in the adaptation of a new technology. First of all, we have a reluctance to use the new technology; we close our eyes and pretend it isn't there. For example, in word processing technology, we may characterize this stage with statements like "I've been using an XYZ brand typewriter for 20 years, and it always works just fine. What are these word processors? I really don't need one, they are just too expensive." This is the *reluctance to use* stage. In the second stage, the *replication* stage, technology attempts to replicate the previous technology. If we return to the typewriter example, we might find IBM replacing its Selectric with a "memory" typewriter, with a single line display and some limited editing capabilities. This uses just a little of the new technology but does not "embrace" it, i.e., fully take advantage of all of its features to do new things. We are still using the old technology, but are simply making it slightly better. We are copying the way typing was always done, i.e., line by line.

In cartography, computer cartography was ignored for some time, as in stage 1, and then was faced by replication, producing pen plotters and table digitizers, simply updated electronic versions of the pens and drafting tables we had always used. The plotters were simply mechanical arms with pens fixed to them, and we still fed them pieces of paper and changed their ink. We took a digitizing tablet, much like a drafting table, and instead of using a mapping pen we used a cursor to draw lines. We replicated the previous technology, making maps exactly the way we used to, using only some aspects of the new technology.

The third stage is the full implementation of the new technology, in which we forget the previous technology, and the new technology becomes the current technology. Cartographers first had a reluctance to use computers altogether ("no computer can draw a map the way I can"), then they replicated the previous technology ("well maybe those computers can draw maps after all, but let me see them draw on Mylar and Leroy stencil the way I can"). Finally, in the full implementation of the technology we have to ask new questions altogether. This is a sign that a revolution has taken place, since new ideas are necessary to organize the new approach. We may ask, for example, if maps actually have to be on paper. New media are now available, such as microfilm, video, broadcast images, and holograms. Perhaps, more simply, we now have to ask ourselves just what a map actually is.

A characteristic of a developing technology is that we move through these stages. Also, however, we find that all stages usually exist at the same time. We still have both an ignorance of and a reluctance to use this technology. We have replication of the previous technology, but fortunately we also have some full implementations, and there are some good examples of people who have adopted computer technology and made a success of it.

1.3 THE HISTORY OF COMPUTER CARTOGRAPHY

Computer cartography in the United States really dates back to a single article written by a graduate student at the University of Washington, Waldo Tobler. The paper "Automation and Cartography" was published in the *Geographical Review* in 1959. At that time, plotting devices were simple cathode ray tubes, and input and output were normally by punched cards. During the early years, the 1960s, the accent was upon the creation of algorithms, i.e., the creation of expressions of ways of doing things mechanically that had previously been done by hand. Since programming these algorithms was difficult, many remained unimplemented. So in the past, contour lines had been drawn by a cartographer, using his knowledge about the lay of the land and his perceptions about how the map should look. Later, the computer gained this ability, giving the cartographer the analytical role of deciding how best to represent the lay of the land rather than how to draw contour lines.

The first problem that had to be solved was how to make a computer draw cartographic lines. The way to do that was to examine how the cartographer had made the decision, what methods were in use, and how they could be automated. In fact, many of those decisions are made on a very simple mathematical basis, and there is often a simple algorithm which will replace the method the cartographer had been using.

Cartographers devised a plethora of different algorithms, at first implemented as stand-alone computer programs, and later consisting of program packages, peaking with the production in 1968 of the SYMAP package at Harvard. The 1970s saw two major changes. First of all, the implementation of new algorithms brought innovations in producing new types of maps that had previously been impossible, certainly computationally, to produce. Within five years, most cartographic techniques were automated, even some complicated methods such as hill shading and cartograms. Computer programs were created which could produce almost all the various types of cartographic representation. The second change during the 1970s was that people began to realize that applications of computer cartography had potential commercial value, just as a whole first generation of graduates from the universities that were specializing in computer cartography provided a small group of trained students for the job market. Many small companies started during this period, employing the new cartographers and using software engineering practices to produce some really effective and cost-competitive cartographic software. Most of this software was made available for those who wished to make new maps or those who wished to replace manual mapping systems. It should also be noted that federal, state, and local government frequently took the initiative in producing mapping software. This phase continues today, and in fact is gaining momentum as the industry matures and as the microcomputer penetrates smaller and smaller drafting-office environments.

Some of the first programs were for the types of computers that were around in the 1970s, large mainframe computers, mostly IBM, and so used existing languages such as FORTRAN and simple proprietary graphics plotting systems such as CALCOMP and Tektronics PLOT-10. In keeping with the times, most applications programs ran in batch mode and frequently used the line printer as the primary display device. A more detailed discussion of specific software packages and computer programs follows in Chapter 3.

1.4 NEW DISPLAY MEDIA

Computer cartography has dictated the development of new types of hardware. At the same time, computers have become much smaller and cheaper; in fact, we now use computers on our laps. Storage ability has improved greatly, bringing down the need to force things to fit into the smallest possible space. We have whole new storage technologies which did not exist 10 years ago; optical disks, video, bubble, etc. A world atlas, for example, can now fit onto a 3-inch video disk.

In Britain, a photographic and cartographic data-base for the entire country is available using CD-ROM technology, appropriately named the Doomsday project (Rhind and Openshaw, 1987). Using this technology, every sheet map in the world could be stored in a very small space. We have new methods and new display media, which have broadened our minds as to what a map is. We used to think of a map as something we could roll up in a tube. Now a map can be a set of electrical impulses on a video tube. In fact, the TV tube is now a viable map distribution system in several countries. Canada, the U.K., and France, for example, use videotex, in which the user can interactively select which broadcast video to see using a keypad much like the channel selector on a remote-control TV unit. Many of the "pages" of information on these systems are maps. We can think of a map independently of the technology or device by which the actual image is to be prepared. Moellering has termed such maps "virtual maps" (Moellering, 1983).

Maps of immediate interest, such as satellite and interpreted weather maps, the locations of accidents and places in the news, travel patterns, and road blockages, can be interactively received having been updated on an hourly basis. These maps never exist other than as sets of electrons falling on phosphors, and we have had to broaden our definition of what a map is to include these new media. Probably the most common map that the average person sees on a daily basis is the TV network news weather map. These maps are mostly digitally produced, with the weather man standing in front of a blank wall and the camera zooming onto a computer screen showing the weather map. The digital technology in this respect is the most commonly used. In addition, the actual printing of maps back onto paper is now possible using new methods such as ink-jet and laser-jet technology, and direct plotting onto film and microfilm.

1.5 NEW CARTOGRAPHIC PROBLEMS

We are moving toward automating even the toughest cartographic problems, among them the automated placement of text, an often ignored but essential part of maps, and increasingly making maps directly from images, especially air photos and satellite images. Dealing with text has been a particularly difficult issue for computer cartography, especially since the production of text in traditional stick-up methods was undergoing its own technological revolution due to phototypesetting, laser printing, and new drafting and reproduction materials. A map is not just a collection of lines, colors, and polygons, it also has important textual information. The selection, placement, and production of text is a very important part of cartography. Previously, this aspect was virtually ignored by almost all computer mapping systems. At last we are dealing not only with some of the issues related to putting text on maps, but we are also working to make the computer select and place text where appropriate. To be able to have the computer decide where to put the text could automate the single most time-consuming task in manual cartography. Text positioning must be such that it does not overlap other labels or important information, and that the laws of text placement are obeyed. A pioneer system for automated text placement is the AUTONAP system (Figure 1.01 on page 6).

Finally, once the text has been selected and placed, the symbolization is critical to the esthetics of the map. Choice of font, color, spacing, and the text slope are variables with which individual cartographers have been able to give a map a certain "style." Much of the "artificial" look of computer-produced maps can be traced to the text design. A really effective computer mapping system would give the cartographer control over this intangible quality of maps.

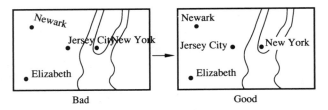

Sample Name Placement Rules:

1. A name must lie totally within the bounds of the map.
2. A name should not overlap with another name or with a feature.
3. If a name overlaps with a line feature, the line, not the name should be interrupted.
4. If a name is to be horizontal, it should be parallel to the rectangular coordinates on a large scale map.

Figure 1.01   Sample Text Placement Rules from the AUTONAP System

Another really sticky problem is making maps directly from images. In the past we used to go out and survey areas using plane tables. Later this was replaced with photogrammetry, using air photos to compile and update maps. Increasingly, satellites are reaching higher and higher resolutions, making them suitable for mapping at larger and larger scales, certainly scales on the order of 1:50,000. How to make maps directly from these images without the intervention of a human interpreter is the final part of this problem. Progress is being made in this area, especially in the fields of remote sensing and image processing. Features like roads and rivers can now be identified from imagery and replaced directly with the appropriate cartographic symbols. For some of these things we are having to use artificial intelligence methods.

Other parallel fields are also influencing current events in computer cartography. Surveying is undergoing its own computer revolution, and many surveyors now use COGO (coordinate geometry) systems, automatic note takers, and pen plotters. Similarly, the computer-aided design (CAD) industry is producing highly sophisticated tools for automating the traditional drafting process, in much the same way that paint programs are automating the artistic painting process. These technologies are influencing and being influenced by developments in computer cartography.

1.6 THE COMPUTER'S INFLUENCE ON CARTOGRAPHY

What have been the implications of these technological changes, and what has happened to mapping? First of all, cartography is now fast, or at least faster. The compilation time for a 7.5-minute quadrangle, with field surveys and checking in the 1940s, when most of the series was made, was on the order of five years. To update the maps now, we use photogrammetric methods, and the survey and field work plus the cartography takes about two years. At the time extreme, we can take a military example. Consider the scenario of a rather aging reconnaissance satellite. If map coverage is required for any particular place, the satellite is repositioned in orbit, photographs are taken, and the film is dropped into a reentry case and ejected. The case enters the earth's atmosphere and is caught by a net towed behind a large plane, is processed, and is delivered to the Pentagon within a couple of hours of having been taken, allowing almost instantaneous updating of maps. Systematic coverage using this type of method is impossible, so for this we use other satellites at coarser resolutions. These systems are not futuristic, having been in orbit for over a decade. The French, Soviets, and Japanese now have advanced mapping sensors in orbit which can give map turnaround of only a few hours. Mapping speed has been increased substantially by computer mapping systems.

The computer has also increased the potential map accuracy, although accuracy is more difficult to discuss, because most people confuse accuracy and precision. If someone were to ask me the time, and I replied that it was 4:12:15.783, I would be very precise. I might be completely wrong — the time could actually be 5:31 — but I would be very precise. If I were to say about half past five then I would be fairly accurate, though imprecise. Has computer mapping made maps more accurate? It has certainly made them more precise. We are able to store and use many more numbers now and we can compute lengths and areas with more significant digits than before. A surveyor can calculate the area of a new subdivision plot to the thousandth of a hectare. Are these numbers more accurate? It certainly seems that we have taken the old causes for inaccuracy, smudgy lines, and shaky hands, and replaced them with other things, such as digitizer's neck and grid interpolation error. I would argue that perhaps in the very long term, computers have made maps more accurate. More important, however, is the fact that our accuracy levels are now measurable against the truth, or at least against other maps, and we have therefore made our maps more accountable. Much recent cartographic research has focused on the accuracy of digital maps, and some important findings have already found their way into cartographic theory.

A major innovation in surveying is the implementation of a system involving 18 geostationary satellites in orbit around the earth, whose locations are very precisely and accurately known, since the orbits are very predictable. The system is called the global positioning system (GPS). This means that if you have a receiver which receives and decodes the signals coming from these satellites, you can fix your location very precisely by three-dimensional triangulation from a set of three satellites. It is not unknown now for a $50,000 system with some microcomputer post-processing to give you latitude, longitude, and elevation to within about 10 millimeters. That is so accurate and precise that it is good enough to measure continental drift; in fact, this has contributed a great deal to us figuring out that the continents are moving around and by how much. The contributions of this satellite system to the mapping sciences of geodesy and surveying are already apparent.

This new technology allows us to resurvey things with a new level of accuracy, and sometimes we have found that we have gotten things wrong, in some cases really wrong. A good example would be the Pacific Islands, totally out of sight of every other piece of land in all directions, so that locations can only be fixed by astronomical measurements of latitude and longitude. With a GPS receiver we have found out that our maps show the islands in some cases tens of kilometers off their correct locations. In many senses, cartography has a long history of islands which appear and reappear, and move around with changes in mapping technology rather than having anything to do with Atlantis, the Bermuda triangle, or plate tectonics. A whole new promontory was discovered in Antarctica, the Washington Monument was found to have been mislocated by several meters, and probably the most famous instance is the fact that Mount Everest spent a short period between measurements as the second-highest rather than the highest mountain on earth.

The digital computer allows us to gather cartographic data which is both precise and accurate. Unfortunately, the implication of quality that comes with accuracy and precision has not always been justified. When the source material for a digital map is a sheet map, the digital map can become a faithful exact reproduction of all the errors involved in putting the original map onto paper. It is important to retain a "fitness for use" criterion when considering the accuracy of digital cartographic data.

1.7 BENEFITS OF THE COMPUTER TO CARTOGRAPHY

What are some of the benefits of using computers in cartography, and has the computer made mapping more cost-effective? It is rather expensive to produce maps, and in the initial stages it was believed that computers would make it incredibly inexpensive to produce maps. In fact, in the long run, computer-produced maps seem to cost about the same, with a few exceptions, as they do to make by hand. On the other hand, it probably will not remain that way. If we think about the stages of adopting a new technology, much of the cost of computer cartography is due to replication of the previous technology. People use figures such as the cost of digitizing a sheet map, normally a one-time fixed cost, with consequent far lower variable costs per map from the single base.

The level of output of maps has increased using computers. Most of the fixed costs of computer cartography are in preparing a new base map, and after this it is both easier and less costly to produce maps from the base. For example, we may spend a great deal of money producing a new census tract base map after every decennial census, but in doing so we are ensuring that we can map on that base any of the variables from the census, and that we can use the map base for numerous other derived maps and analyses. Only the data and the symbolization and map types change. This implies that we can increase output. What this means is that mapping projects which were previously not possible are now cost-effective, even cheap, perhaps affordable by third-world countries and local governments which were previously excluded by cost from the mapping business. It could be argued that these applications have the greatest potential for computer cartography, for they have no previous technology to waste time replicating and can reap the benefits of other's experience.

The overall effects of the computer on the discipline of cartography are many. Cartography in the nineteen fifties was much of a service discipline. A cartographer would be attached to a geography or geology program academically, or isolated in a map department commercially. A cartographer's function was as a technician, a producer of maps, rather than of systems for making maps. Wolter (1975) argued in *The Emerging Discipline of Cartography* that cartography has evolved through its self-examination as a result of the computer age all of the requirements to become a distinct academic discipline in its own right. Certainly, if one examines the literature, this is evident. Much that is published in cartographic journals these days is clearly analytical cartography, although there are other approaches to the discipline. One could argue that this is just a return to the preeminence of cartography, a return to the role that cartography held under the ancient Greeks or during the age of discovery. Many more people outside the discipline now encounter and eventually study cartography. As a result of this scrutiny from outside, we have been forced to define the cartographer's domain of interest very precisely. We have had to define in detail exactly what cartographers do, how they do it, exactly what the information they deal with is, what its limitations are, and have had to model the flow of this information through the mapping process.

The computer has relieved the cartographer from tedious production tasks. Anyone who has experience with manual cartography knows how much hard work has to go into a manually produced map, even one which is poorly designed or inexpertly drafted. The focus of cartography used to be on the production technology. If the computer is doing the minor production tasks for us, we can worry about other things, for example, design — ultimately what a cartographer is trying to perfect. For one thing, the computer has allowed us to set up a design loop. The manual equivalent is to work on a series of separations one at a time, incorporating a single map design for each map project. We finally get to producing a map proof, and for the first time after perhaps 100 hours of work, get to see what the map looks like. What if we don't like what we see? More commonly, we might say "if only this text were slightly smaller" or "if only this red were green." It is too late to change the map; in other words, we have no design loop.

In a design loop, the finished product can be interactively modified to make all the changes we may wish to incorporate. In fact, we can use the design loop to produce experimental maps which will help us to learn better design: for example, to see why a map with 12 fonts and 30 shade classes really does look bad. This changes the emphasis from the technology to the design, and further to the principles which make this a good design regardless of the technology. In this way, we can make maps look better — not a bad definition of what it is that cartographers do.

The computer has produced a whole series of new capabilities. The ability to manipulate color is a good example of this. The computer has given us new types of maps and new media for their display. The Museum of Holography in New York, for example, has several cartographic exhibits. These are maps that can be walked around and perceived as three-dimensional objects. Similarly, we can use cartographic techniques to map new types of data, such as statistical distributions, the ocean floor, Mars, or a molecule.

1.8 DISADVANTAGES OF THE COMPUTER FOR CARTOGRAPHY

The computer has had some negative effects on cartography. For one thing, the amount of technical training that a cartographer has to acquire has increased enormously. The cartographer of the nineteen nineties must be a data-base expert, a user-interface designer, a software engineer, retain a sense of map esthetics, and still produce maps.

Cartographers ideally need training in image processing, remote sensing, photogrammetry, land information systems, geographic information systems, surveying and geodesy, and computer programming.

Another disadvantage of the computer for cartography as a discipline is that those not trained in cartography can now easily produce maps. While this is in many cases a popularization of the mapping process, in the early days of computer cartography this led to a renewed need for good design and esthetics. For a while, replicating the previous technology, mapmakers threw out the quality standards which that technology had contributed to cartography as a whole in the analytical sense. For about 20 years computer cartography produced rather inferior products, which were accepted not because they were better but because they were new and different. At last, we are finally producing maps by computer that are as good or better esthetically than those we can make by hand.

1.9 SOME GENERAL TRENDS

There have been many general trends within cartography as a result of the advent of the computer. One of the trends has been towards simpler maps. If we were stuck with a single product, we often believed that there were savings to be made by incorporating as much data as possible on a single map. Maps have become "single message," in much the same way as business graphics, with a good example being the weather map or the sales map. These are maps being used to communicate a message. A whole school of thought within cartography has concentrated upon the effectiveness of maps in communicating this message to the map reader. This communication school has sought to improve map design to get the message across faster and more effectively.

The computer has given us defensible design. We can assign numbers to back up qualitative judgements. A good example is to use recognized error reduction techniques to choose class breaks for shaded maps, or known methods to do line generalization and simplification. So an island may be eliminated on a map at a certain scale because it falls below a certain threshold area, rather than because it was easy to paint over the island with opaquing fluid on a negative.

The computer has promoted new types of cartographic symbolization and representation; i.e., we have devised new types of maps as well as new ways of making the old types. A good example is the depiction of terrain. Now we can generate realistic depictions of terrain to simulate the way the terrain will actually look to an observer. In particular, the computer has made thematic mapping available to those who are not cartographers but who work with statistical data with a geographic component. The computer, therefore, has opened up mapping to those without a formal training in the discipline. This may yet prove to be the most critical advance in the acceptance of computer mapping.

Finally, education has expanded cartography to include computer and analytical aspects. We have had an enormous increase in both the numbers of, and demand for, the type of college course at which this book is aimed. Almost every geography program in the country offers at least some form of computer (not necessarily analytical) cartography. Sometimes schools offer multi-semester sequences, even entire specializations in computer cartography. Increasingly, the graduates of these classes go on to find employment in the booming industry of computer cartography, hopefully well endowed with a sound base in the analytical aspects of the discipline which will ensure those same students of employment in the future.

1.10 GEOGRAPHIC INFORMATION SYSTEMS

The influence of the computer on cartography has opened up cartographic methods and techniques to more general purpose uses of geographic information. Nowhere has this been more evident than in the field of geographic information systems (GIS). GISs are automated systems for the capture, storage, retrieval, analysis, and display of spatial data. Large numbers of these systems are now available to assist resource and spatial data managers with their planning and decision making, as well as routine record keeping and inventory. Since these systems use geographic data extensively, most of the queries made of these systems provide solutions which are cartographic. As a result, these systems are usually supplied with either an interface to a separate cartographic display system, or contain their own such systems. The type, quality, and capabilities of these display systems vary remarkably from system to system, going from poor to the very best in quality. The major differences in how GISs generate maps are twofold. First, the map is either an intermediate or partial solution to a query, and may therefore not contain any of the true cartographic elements other than a figure and a georeference system. This is not to say that the final maps which provide the problem solutions are not good cartography, but reflects instead the emphasis on query and response rather than display.

Second, the user of a GIS is not always a cartographer, and therefore does not place the same level of importance on the map as on the message the map conveys. GISs have become the primary way in which people are first introduced to computer cartography, and in some cases to analytical cartography. It is important to realize, however, that the goals of computer cartography and the goals of geographic information processing are not always identical, though they are always similar. The skilled cartographer should develop an understanding of GIS, and should see in them the challenge of a new set of demands upon the theory and methods of cartography. It is not enough to state that the computer cartographer is interested solely in the graphic display of digital cartographic data, since the structuring, encoding, and the representation of the data are the very basis upon which analytical cartography is built. Cartographic transformations, the basis of analytical cartography, underlie all geographic data processing, and therefore are of overlapping interest to cartographers and the builders and users of geographic information systems alike.

1.11 REFERENCES

MOELLERING, H. (1983) "Designing interactive cartographic systems using the concepts of real and virtual maps," *Proceedings, AUTOCARTO 6*, Sixth International Symposium on Computer-Assisted Cartography, Ottawa, Ontario, October 16-21, vol. 2, pp. 53-64.

MORRISON, J. L. (1980) "Computer technology and cartographic change," in D. R. F. Taylor (Ed.), *The Computer in Contemporary Cartography*, Wiley, New York, chapter 2, pp. 5-23.

RHIND, D., AND S. OPENSHAW, (1987) "The BBC Doomsday system: a nation-wide GIS for $4448", *Proceedings, AUTOCARTO 8*, Eighth International Symposium on Computer-Assisted Cartography, Baltimore, MD, March 29-April 3, pp. 595-603.

TOBLER, W. R. (1959) "Automation and cartography," *Geographical Review*, vol. 49, pp. 526-534.

TOBLER, W. R. (1976) "Analytical cartography," *The American Cartographer*, vol. 3, no. 1, pp. 21-31.

WOLTER, J. A. (1975) *The emerging discipline of cartography*, Department of Geography, University of Minnesota, Ph.D. Dissertation, University Microfilms, Ann Arbor, MI.

2

Hardware

2.1 DEVICES FOR COMPUTER CARTOGRAPHY

2.1.1 A Map of the Computer

In Chapter 1 we saw that the impact of the digital computer upon cartography has been revolutionary. To understand the tools of contemporary cartography, therefore, we must examine computers in depth. We shall begin by considering the nuts and bolts of computing, the hardware. This will allow us to focus on the special types of devices required for computer cartography, which are often unlike those required for more routine computer tasks. To conclude the chapter, the relatively recent addition of the workstation will be introduced, and its significance for both analytical and computer cartography demonstrated.

Hardware consists of a set of physical devices which perform electrical, mechanical, or other physical tasks or state changes according to instructions given by software. Hardware is something breakable, needing mechanical assistance, repair, or maintenance, and reflects a particular stage of technology, therefore implying eventual obsolescence. The center of a computer is therefore of relatively minor importance for analytical cartography, though of some importance for computer cartography. Of more importance to cartography is the peripheral device, a plug-in or add-on part of the computer which gives the computer cartographic capabilities. A recent trend has seen peripheral devices with significant amounts of computing power, sometimes more than the "main" computer itself.

Figure 2.01 presents a *map of the computer*. At the heart of the computer is the central processing unit (CPU). The CPU consists of the part of the computer which actually carries out the instructions given by software, such as the addition of values in

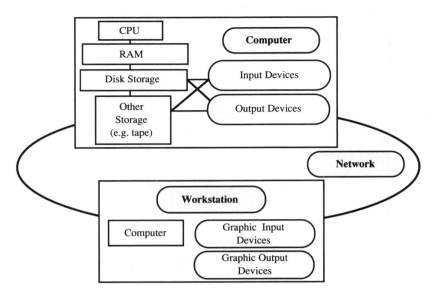

Figure 2.01   A Map of the Computer

registers. In a microcomputer, the CPU can be as small as a single microcircuit or "chip." CPUs vary in power by the size of their registers in bits, and by their speeds, usually measured by both their clock speeds in megahertz (millions of instructions per second), and by their ability to handle complex operations (megaflops, or millions of floating point operations per second). Many CPUs are enhanced with an accelerator, which in particular increases the speed at which the CPU can handle floating point arithmetic. Most computers can do integer arithmetic at high speeds, but slow down considerably when the use of floating point causes the registers to overflow.

The memory locations accessed by the CPU have to match the speed at which the CPU operates and therefore are limited in size. In addition, this memory must be able to store instructions, data, and parts of the operating system. Such "fast" memory must be accessible starting at any point, and is therefore known as random access memory (RAM). Even large computers may have fairly small RAMs, and most minicomputers have about 4 megabytes or less. While the first microcomputers were limited by the size of the link to the CPU (the bus) to about 28 kilobytes or less, most microcomputers now have over 640 kilobytes, with some well over a megabyte. Since the size of the RAM is the upper limit on the address space available to computer programs, this is often a limiting factor for cartographic software.

Data and instructions awaiting the services of the RAM and the CPU reside in some kind of permanent or temporary storage. In large computers, many different kinds of storage are available. Among the more temporary storage mechanisms are fixed and removable disks. Fixed disks are installed within the computer and involve a mechanical device capable of writing information onto and retrieving it from a storage medium.

Magnetic platters, optical disks, and even "bubble" memory are used to store information. Removable systems allow data and programs to be archived or added to the system. Removable systems include disk cartridges, magnetic tapes, floppy disks, and removable optical media. Most cartographic software and data are available on micro-disk or magnetic tape. Among the disks, the 133 millimeter (5.25 inch) and the 90 millimeter (3.5 inch), are standard. Among magnetic tapes, the nine-track 13 millimeter (0.5 inch) tape at the 1600- and 6250-bit per inch densities are most common. These media are comparatively cheap, widely available, and are reliable enough to store data without error for several years.

Computer cartography involves interaction with the computer in ways which are more complex than the simple video terminal and keyboard which microcomputers have made household features. In particular, we need to enter into the computer graphic locational information and to get from the computer output in the form of maps. This involves adding to the computer devices which are specially made for graphics. Such add-on features are called peripherals, although this term includes add-on memory, disks, keyboards, mice, etc. A more specific term for a peripheral is a *device*, and the devices we require for cartography are graphic input and output devices. This chapter is, therefore, largely a summary of the various graphic input and output devices which have been used for computer cartography. Historically, these devices were used in isolation, connected remotely to a large computer, performing their functions one at a time, so that their use in cartography was much the same as when maps were produced manually. More recent developments, however, have reshaped the way devices are used. This development has been made possible by two developments, the development of computer networks and the development of the workstation.

Computer networks are groups of linked computers which can share the various devices connected to the network rather than to individual computers. Within a network, each CPU, with its RAM, becomes a client, and is connected to the network via a cable and network software, which handles the movement of data through the network and manages the individual systems. Peripheral devices, such as disk storage, tape drives, and especially input and output devices, can then be connected to the network rather than simply to an isolated CPU. One of the CPUs on the network typically functions as the master, usually supplying most of the disk storage and services. This CPU is known as the host, and if it supplies the disk space to the network, is known as the file server. In the case of computer cartographic software, installation is often on the file server. The software is then "released" to all of the users of the network.

Networks can be extended beyond the limits of one system by telephone and other links to other computers. Thus entire data-bases can be distributed, resident on different computers around the nation, or even the world, and used by clients remotely. Such networks are available even to the smallest system, including microcomputers, assuming the availability of a modem.

The second important development for cartography is the graphic workstation. In the most primitive form, a system to produce graphics contains only additional RAM connected to a graphic display card, and a monitor capable of supporting graphic display. A workstation, however, can support any number of graphic input and output devices simultaneously. As a minimum, the workstation should have both a graphics and a text display, and some sort of graphic input device, such as a digitizer, mouse, or touchscreen. Many workstations incorporate more sophisticated devices. A workstation has its own operating environment and is usually connected to a network, where data and additional software are available.

2.1.2 Input Devices

In the preceding section we met the concept of a peripheral device. In computer cartography, the devices of major concern can be classified as input, those which provide data, and output, those which produce maps or other information. Within this division, we can break the devices down further into map and nonmap.

Non-map input devices, though in regular use, are not particularly cartographic. Examples are the keyboard, a computer tape, or a disk from which a file can be read. Anything that can send data to a computer qualifies as an input device. Map input devices are many and varied. The most primitive, and first, was the up, down, left, and right arrow keys on a keyboard, sometimes mapped onto the h, j, k, and l keys. Continuing chronologically, the next was the joystick. Joysticks were used occasionally in computer cartography, but remain in use solely for games. Some graphics terminals included on the keyboards two wheels which moved a set of cross-hairs on the screen in the x and y directions. Positioning the cross-hairs and then touching a key returned the screen location to the device. This type of input has largely been replaced with the light pen, the digitizing pad, the touch screen, the track ball, and the mouse.

The light pen is favored in CAD applications, where interactive screen modification of a picture is important. The light pen is much like a pen, with the exception of the fact that the point end emits a light signal which is detected by the screen, and the other trails a wire. The digitizing pad is a small digitizing tablet, sometimes only about 200 millimeters on a side. Usually, the tablet includes, or has attached to it, a menu, from which commands can be selected by pointing to the menu area indicted and either pressing a button on a cursor, or in the case of a pen cursor, pushing down on the pen tip. The graphic interaction takes place away from the screen, although when interactive graphics use the screen, the cursor position is usually continuously displayed on the screen. Many "paint" packages, the digital equivalent of a pen and paintbrush, use the digitizer pad. Figure 2.02 shows an interactive graphics system which supports input from a joystick, a trackball, and a small digitizing pad with a cursor.

Another form of graphic interaction is the touch screen. On a touch screen, the screen is partitioned into menu areas, on which is displayed text or graphics describing the selection. When the user touches the screen, the touch is detected and the software reacts accordingly. Touch screens have found their way into supermarket guides, hotel and airport lobbies, and libraries, where searches are a set of hierarchically linked choices, narrowing in on a final image to be displayed. In many cases, these systems include maps. For example, a hotel lobby screen may contain a neighborhood map with restaurants which can be queried by the ethnicity of the food served.

Figure 2.02   Graphics System with Multiple Interactive Input Capabilities (Photo permission of Chromatics, Inc)

The track ball is a ball mounted in a holder which can be moved freely in any direction using the palm of the hand. Often buttons or dials allow the motion to be constrained to one direction, or to proceed just one screen unit at a time. As the ball is moved, so the screen cursor moves with the same relative movements as the ball. Most track balls use "wraparound"; that is, when the cursor is moved off the screen to the top, it reappears at the bottom, etc.

By far the most common graphics input device is the mouse. Two types are in use: the mechanical and the optical. The mechanical mouse has a small plastic or rubber ball inside which is turned by friction against the table top or a special mouse pad as the mouse is moved. The relative movements are picked up by the computer, and are used to control the motion of a cursor on the screen. The optical mouse uses a reflective pad and a light beam to detect motion, which avoids the problem of dirt and dust clogging the ball in a mechanical mouse. Recently, cordless mice have appeared, which allow complete freedom of movement since they need no wire connection to the keyboard. Using a mouse involves first moving the cursor around, but also selecting features from a graphic menu. Items are selected by clicking, that is, moving the cursor into the correct location and pressing a button on the mouse. Holding down a button sometimes reveals another level of menu, the so-called pull-down menus used in many graphics-based operating systems, which require more cursor movements. Selecting a feature by holding down a button while the cursor is over a feature, and then moving the feature by moving the mouse while holding down the selection button, is called dragging. Dragging is an integral part of using CAD and draw-type software.

The map input devices covered so far have in common the fact that while they have some graphic capabilities, the locational data they provide is too approximate for anything other than selection from menus. These devices support the graphics function of selection, that is, indicating approximate direction and location, and picking, the choosing of a menu object. Detailed cartographic input comes from one of two sources, semi-automatic digitizing and automatic digitizing.

Semi-automatic digitizers are the computer equivalent of the drafting table. They consist of a tablet, which can be as small as a printed page, or as large as 1.5 by 2 meters. Smaller tablets can rest on a desk top, but the larger ones require stands and include options such as power movement and back lighting. While a number of different options are available for the cursor on a digitizer, most cursors have between 1 and 16 buttons, and have options such as magnifying lenses, choices of bull's-eye and cross-hair patterns, and the ability to follow lines within the cursor window automatically. Figure 2.03 shows a typical array of tablet and cursor choices. Cursors with multiple buttons allow data such as elevations and labels to be entered along with the map information, which means that the user need not move between the keyboard and the tablet.

Generally, the tablet itself consists of wires embedded in the surface along the cartesian axes. Each wire is connected along the edges, so that when the cursor is moved to a point and the button is pressed, an electrical charge is generated by a coil in the cursor which is detected on one of the wires underneath. The unique wire which picked up the charge gives the X and Y values. The signal is then converted to a string of ASCII characters corresponding to the button pressed, the x and y values of the point, and any attributes entered. This data is sent down a cable to the controlling computer, usually to a microcomputer. Early systems often converted the signal to codes for transmission over telephone lines, since much early digitizing took place with remote connections to large computers or directly to magnetic tape off-line. This arrangement has been replaced so that the signal from the tablet now moves straight to a special purpose board inside a microcomputer which controls the tablet. In this case, the data moves directly to microdisk or hard disk.

Electronics is not the only technology that has been used in digitizing. For a long while there were sonic digitizing tablets, still seen occasionally. Sonic tablets used a drafting table, but instead of wires set into the tablet there was a string of microphones along the top and one side. The button on the cursor simply emits a click or a beep, which is picked up by the microphones. A triangulation was then performed to figure

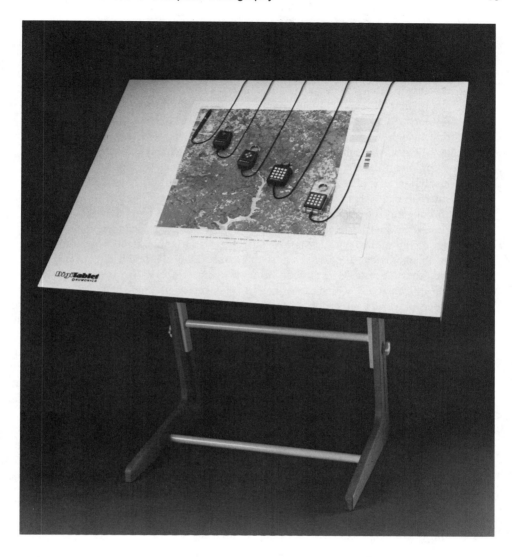

Figure 2.03   Digitizing Tablet (Photo courtesy of Numonics)

out where the cursor was. Other tablets are simply mechanical arms like pantographs, which use the mechanical movement of the arms to compute x and y. These systems are usually inexpensive, and can be mounted on an existing table.

Many digitizing tablets support voice data input in addition to the cursor. Simple commands can be spoken, and a microphone on the controlling computer matches the recorded voice against a preset list of commands. This allows much more efficient use of the tablet, since the users attention remains on the task, not on moving between menus and cursor buttons. Also in use are three-dimensional digitizers, where a cursor on a cantilever is moved around an object touching the edges to record three locational dimensions. These systems have been used little in cartography.

Semi-automated digitizers are the most common when maps are concerned, and the larger sizes are required for digitizing sheet maps. The alternative technology, automatic digitizing, is more recent. Automatic digitizers include scanners and line followers. At the low-cost end of the scanner market, two types of scanner exist. The first, which were designed to scan pages of text and automatically convert the pages to text files in the computer, are not suitable for cartography, with the possible exception of producing text lists for typesetting. Some of these scanners are capable of low-resolution graphics, and parts of small maps can be scanned into paint and draw software using this technology. Most of these scanners operate in pixel mode in monochrome, but some conversion to vector or object mode is now possible.

The alternative scanning technology is the video scanner. The video scanner is the digital equivalent of a television camera and is mounted on a stand. Maps, images, or air photos can be placed under the camera, and the television picture is shown on a monitor and rasterized, turned into a series of numbers. An identical technology exists for still cameras, which store their images on small disks and replay them through a video player as television images. Normally, what is returned is exactly what comes back from a remote sensing device, the gray tone or the intensity in a certain light frequency for each little square area or pixel on the image. Red, green and blue channels for a color map can be scanned separately and resuperimposed to generate color from three gray-scale images. Geocoding on these images is usually established using reference or control points and rubber sheeting, topics covered in later chapters. Only small areas of maps can be scanned with these low-cost systems with true cartographic fidelity, though the resolution capabilities of these systems are improving rapidly.

The top-end scanners are of very high resolution and as a result are expensive. High-precision scanning devices use what is called a flying spot technique, which involves moving a light or laser beam very quickly across the surface and detecting either absorption or reflectance at every point on the surface. Since paper maps can fold, fade, or tear, the source map is usually the film separation from which the map was printed. Monochrome scanners can be used to scan each ink-color separation separately for any map. Increasingly, however, color scanners are capable of extracting colored features directly.

One of the problems with scanning maps is that most maps have a great deal of redundancy. Polygon maps contain information only at the boundaries. Also, text is not easily processed from scanned data. Other problems mean that much post-processing must be conducted on scanned maps. For example, the map separation containing contour lines can be scanned simply, yet the breaks in the lines for contour labels must be joined together after scanning to give continuity, and the labels deleted and turned into an accompanying attribute file. Also, scanners often have resolutions which supply an immense amount of data for very small pixels. As will be seen in the second section of this book, geocoding maps involves far more than simply converting a map into sets of numbers.

Scanners are in widespread use for computer cartography. Many of the earlier problems with scanning have been eliminated. An important breakthrough was the development of hardware for the automatic conversion of the scanned data to strings of (x,y) pairs, known as vector data. These special purpose vectorizers have made scanning a valid alternative for mass digitizing. Scanning is already widely used in high-quality production cartography. As the costs and capabilities of scanner evolve, it is highly likely that they will assume most, if not all, of the map input process in computer cartography.

2.1.3 Output Devices

As before, we can divide output devices into map and non-map. Good examples are such things as the line printer, disk, tape, and any other non-graphic medium. The line printer was used as an output device by the very first generation of automated mapping systems. The limited capabilities of the line printer were augmented by overstriking characters to get gray tones, by printing symbols, and by retouching the final output.

Contemporary map output devices fall into several types. The first type are plotters. Plotters were an example of how computer cartography tried to replicate the previous technology. Plotters do this by drawing a map on paper. If digitizing can be thought of as an analog-to-digital conversion, then plotters convert the digital back to analog information. Analog (especially paper) maps have some advantages for storage, and are convenient to carry, especially into the field.

Pen plotters fall into two different types, flatbed and drum. For many years flatbed plotters were cheap, readily available but not really of high cartographic quality. Drum plotters were expensive, difficult to operate, but produced much higher precision, better-quality output. Now, however, some extremely inexpensive drum plotters are available, while some of the very highest quality cartographic output is plotted, or more usually scribed, by a flatbed plotter.

Both flatbed and drum plotters are vector devices. Flatbed plotters take a sheet of paper, film, or scribecoat and attach it to a flat surface. The pen then moves over the paper, which is held in place with tape or more frequently by an electrostatic field which holds the paper perfectly flat. The plotter then simply moves a pen along two axes. Different technologies determine the actual drawing. The pens can move on an arm, which itself moves along an axis; the flat bed can move in one direction and the pen in the other; or any combination of these movements is possible. All plotters have only two fundamental basic operations: move with the pen down, and move with the pen up. If you move around with the pen down, it draws. With a flatbed plotter you typically have to load each sheet of paper or film yourself and take it out again when it is finished. You also have to load the pens yourself, although even the cheapest plotters are capable of changing pens automatically. More sophisticated plotters can recap pens and change ink without operator assistance. Also available are automatic paper feed and the cutting of plots when complete. This allows the use of rolled paper and the queuing of multiple plots for plotting when the operator is absent.

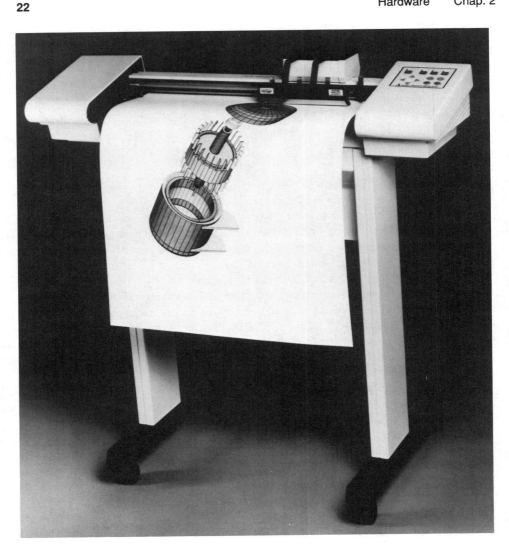

Figure 2.04   A Drum Plotter (Photo Permission of Houston Instruments, Austin, TX)

The drum plotter (Figure 2.04) has a rotating drum with prongs which feed holed paper from a roll. Drum plotters have hoppers which hold one or more pens to support different colors. Motion in the x direction is achieved by moving the paper with the drum past the pens. Motion in the y direction is achieved by moving the hopper across the drum (Figure 2.05). Most drum plotters can use a large variety of pens and plot media. Pencil point, fiber point, rollerball, and refillable ink mapping pens can be used to plot onto paper, film, Mylar, or acetate. Some very high quality maps can result.

Both flatbed and drum plotters, however, are poor at filling areas on maps with continuous color, since the pen has to be dragged backwards and forwards until the area is filled. When all that is required is text and linework, the quality is of cartographic standard.

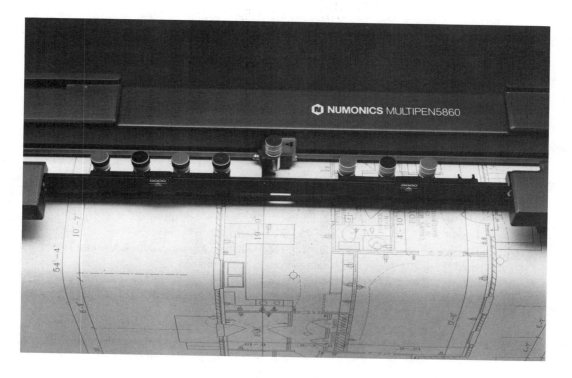

Figure 2.05   Pen Hopper on a Drum Plotter (Photo Permission of Numonics Corp., Montgomery, PA)

Another type of plotter is the electrostatic plotter. Probably the major uses of electrostatic plotters are in the electronics industry, where they are used for printing multi-color circuit boards since they can print a large area of coverage with a great deal of detail. Electrostatic plotters use electrical charges to attract ink to the surface of treated paper. Although electrostatic plotters are bulky, expensive, difficult to maintain, and often support only a few colors, they are better than pen plotters at areal coverage.

Moderately closely related to the electrostatic plotters are printing technologies. Printers use impact through an ink source onto the surface to produce a plot as in a typewriter. The first printers of near-cartographic quality were dot matrix printers. More recently, laser-jet printers have taken over completely. Laser-jet printers have high definition, usually as simply black on white, and have revolutionized desktop publishing and word processing. The resolutions of low-end laser-jets is several hundred dots per inch, but these printers are fully consistent with typesetting-quality printers with over 1,000 dots per inch. Many pieces of cartographic software use the laser-jet as the primary, or even the only, graphic output device. The PostScript standard, widely used by laser-jet printers, has allowed a great deal of device independence, and many drivers for graphics allow maps to be sent to laser-jet plotters. The laser-jet is really a refinement of the xerographic process. The link between publishing, cartography, and typography that this technology offers is another opportunity for cartographers to make superior maps.

Impact printers are capable of low-quality color. However, better-quality color is available without photography using a hybrid technology halfway between dot matrix and laser-jet plotters. Ink-jet plotters are a hybrid because they are not printers and they are not plotters. They use a method where ink is sprayed directly onto the surface. They normally have four ink guns with inks for the production of primary colors of magenta, cyan, and yellow. The fourth gun contains black. Mixtures of these four inks give a very large number of different color possibilities. High-end ink-jet plotters produce almost photographic quality and are getting better all the time. Low-end ink-jet plotters, however, produce rather garish color, can print only fairly narrow strips of paper, and are of low resolution.

The last major category of map output devices are those involving photography. At the high end of film technology is the film writer, which typically writes onto a very large piece of film and thus is favored for the hard-copy of remotely sensed images. At the middle of the range is the graphics camera. Graphics cameras are separate devices connected to the red, green, and blue color channels normally sent from a graphics system to a color monitor. The color image is then reconstructed on a flat display at the end of a box, in the focal plane of a bolt-on camera back. Typically, the exposure is controlled completely by a microprocessor, including the aperture, the light intensities, and the exposure time. Each of the red, green, and blue channels is exposed separately, making the color image additive. Graphics cameras usually support most photographic media, including 35 mm, the various instant picture formats, and various sizes of sheet film. The attraction of this method of hard-copy is the price, which can be very reasonable per image, and the fact that the images are virtually distortion free.

At the low end of the camera market are screen copiers, some of which are very inexpensive. These devices mount onto a black screened box which fits over a monitor. The mount supports usually an instant camera type of camera back, although some support 35 mm film. A number of different processes can be used, including one type of slide film which can be developed in only a few minutes with a stand-alone slide developer. The quality of this type of output is sometimes very good, although care must be taken to avoid the "barrel effect" of cameras focused too close onto a surface which is actually curved and not flat. True flat screens, at first only as liquid crystal displays (LCDs) but now in color and as back-lit plasma screens, are fairly new, but offer distinct advantages to cartography. Last but not least, an inexpensive way to photograph computer maps is to focus a 35 mm camera with a telephoto lens and a tripod onto the screen from a distance of about 2 meters in a darkened room. An f4 and about 1 second for ASA 64 film is required, since the scan lines are caught in the image at speeds greater than about 1/15 of a second. Any reflections or screen edges in the image can be masked out on the slide with tape.

As a final example, some plotters now plot directly onto microfiche and microfilm. This may be one of the major ways that analog maps will be distributed in the future, especially if a hand-held viewer is used. The international report to the ICA published in the *American Cartographer* for 1987, includes two microfiche maps submitted as the best examples of American cartography in the last year. Digital maps can be produced directly from plotting devices onto microfiche or microfilm with no paper involved. These media take up little storage space, and can be mailed in a single envelope instead of a map tube, eliminating rolling and folding.

2.1.4 Vector and Raster Display Technology

A major historical distinction between different computer devices for computer cartography is that between raster and vector technology. These two different display technologies have had a particular influence on much of the theory and data structures involved in analytical and computer cartography. It is worth describing the alternative technologies in a developmental context, to introduce the newer technology which has eliminated many of the inconsistencies between the two technologies.

In raster technology, the screen or the input device is divided into discrete pieces called pixels, which have a finite size. For an orbiting scanner on a spacecraft these would be single areas on the earth's surface for which we can resolve data. On a map they could be latitude and longitude cells or meters marked off on a UTM grid. Each square or rectangular pixel has an attribute associated with it, which could be a color, a light intensity, a gray tone, or a red, green, blue value for that pixel. For input they would be associated with readings or measurements. In terms of output the data could be displayed as is, each pixel mapped onto a piece of the screen.

The alternative is vector technology, in which control of the output device comes from software, which moves a point light or pen across the screen or paper, leaving a mark. Vector input is particularly suited to maps that have lines on them, such as coastlines or rivers, political districts, contours, etc. Raster input is particularly suited for maps that have areal characteristics such as land use, land cover, surface moisture or ocean temperature: an attribute measured over a large area. Vector and raster, therefore, are two different structures, two different technologies for both input and output.

Vector graphics dominated the computer graphics market very early on, with storage tube technology terminals. A storage tube contains an electron gun which sends a continuous stream of electrons toward a screen treated with different phosphors which glow when struck with electrons. Control over movement of the beam is achieved using a series of magnets which allow the beam to be deflected and moved continuously. Since the phosphors store the charge when struck by electrons, the screen stays glowing when hit, and lines, points, and areas can be traced out with the electron beam. When complete, the entire screen has to be cleared at once, usually by depressing a button on the keyboard. Storage tubes allow straight and curved lines to be traced out, although they are poor at filling areas. In addition, virtually all storage tubes are mono-chrome, usually a light green glow on a dark green background.

The advantage of vector displays is that we can draw a line in any direction with any length and have it appear exactly the way it would on a paper map. The finite limit on resolution is the size of the grid of phosphors, a very small spot on the screen sur-face. The disadvantages are several. First, the devices have to be fairly large. My first graphics terminal, a storage tube device, was so large that it had to have its own stand. The display tube on the terminal had to be quite deep and the screen area, though flat, was very small. In addition, the phosphors eventually decayed, leaving images "burned" into the screen. Storage tubes also could be quite slow in drawing a map, although plotting time was a function of the power of the computer behind the terminal. Since selected update was not possible, waiting for redraws could be tedious.

Raster technology now completely dominates the computer graphics industry to such an extent that software when required usually simulates a vector display device using a raster display. Raster technology evolved out of television technology. The screen phosphors in this case respond differentially to electrons by changing red, green, or blue in different intensities. A major difference is that the phosphors are arrayed in a grid as pixels. A look at a color television screen using a magnifying glass will reveal the cells that can illuminate different colors. With a very large number of pixels, they appear to give a continuous image, although in fact it is a discrete image.

In raster displays, the electron beam scans the screen with about one full scan every 1/60 of a second, refreshing or temporarily illuminating each in turn. To actually see the screen redrawing we have to use something that is fast enough so that it sees it only for a very short space of time. A good way to do that is to use your 35 mm cam-era and take a screen photograph. At 1/30 of a second, bands are visible in screen pho-tographs across the picture or the TV screen. This is because while the shutter was open the screen refreshed once and got part way through refreshing a second time before the shutter closed.

The conversion from raster data (preferable for scanning) to vector data (often preferable for plotting) has started to be performed by hardware rather than by software. In Chapter 10, the vector-to-raster (and vice versa) transformations are considered as a software problem. The same task, however, is commonly performed by special purpose devices. One such system is shown in Figure 2.06. Since the vectorization often intro-duces ambiguities, especially in the topology of line and polygon maps, these devices must support extensive operator control over editing. It seems likely that human interac-tion will be used in this editing and matching process for some time to come.

2.1.5 Real and Virtual Maps

With so many different pieces of hardware, capable of producing so many different types of digital map products, it is not inappropriate to ask the question "What is a map?" To answer part of the question, Moellering (1980) introduced the concept of real and virtual maps. A real map is a map which has gone through the symbolization transformation and has found a realization as a tangible object. This implies that the cartographer selected a scale and plotted a specific map type, legend, and text. A virtual

Figure 2.06   Special Purpose Raster-to-Vector Converter (Photo Courtesy of Laser-Scan Laboratories Ltd., Cambridge, England)

map is like a possible map; for example, if we have a digital outline map of the United States and access to state statistical data, then any variable can be mapped, using a number of different methods, classifications, etc. The set of possible maps is finite and even directly accessible using an interactive mapping package. A virtual map, therefore, is both a non-permanent map in itself and also a digital equivalent of a "proof" map, made to reveal the effects of a particular color combination or cartographic technique.

In producing a virtual map we need not be concerned with the specifics of the technology with which the real map will be produced; it need not even have been invented yet. A similar idea is the distinction between hard-copy and soft-copy. First, hard-copy means a map that we have produced that is actually going to last, using a stable medium in a form designed for viewing and use. Soft-copy maps arrived with computer cartography. A soft-copy map corresponds to the work sheet in manual cartography. It is a map we can take a quick look at to see whether we like it or not, whether or not we want to produce a hard-copy at all. With many geographic queries, a soft-copy map can answer our cartographic questions completely, meaning that we never need to produce a hard-copy map.

Softcopy and virtual maps have given computer cartographers the option of using a design loop, a potentially revolutionary tool for improving maps. The design loop is a system by which a map design is improved by interactive trial and error rather than by textbook learning. In manual cartography, map data are compiled, scales and projections are chosen, a set of map types and symbolization techniques are selected, and finally, the cartographer invests the necessary hours in actually producing the map. With so much time invested, one is less likely to see flaws in the map design. If somebody came up and said "Title too large and could be moved over ten millimeters" there is nothing to be done, except to take note for the next time. If the map were a soft-copy, the cartographer could probably make the change in a matter of seconds and then ask the critic for any additional feedback. The ability to change design interactively in this way is the design loop. The design loop also frees the cartographer to learn from the deliberate mistake, and allows experiment with non-traditional designs.

Some soft-copy is indeed soft in that it is actually a temporary image like a FAX. Thermal printers have been used to produce map images, but are usually used only as a cheap intermediate stage before another form of hard-copy is generated. Most paper maps are in reality only longer-lasting versions of thermal prints. Paper tears, expands and shrinks, creases, fades, stains, and is combustible. Virtual maps, since they are technology independent, are virtually indestructible. Unfortunately, they are not yet practical to take hiking in the rain.

2.2 WORKSTATIONS FOR CARTOGRAPHY

The most important hardware revolution for cartography in recent years has been the incorporation of microprocessors directly into the display terminal. In the early days of computer cartography, only large computers did computer cartography, and users of the large computers were remotely connected via dial-in lines and stand-alone "dumb" graphics terminals. Special purpose hardware devices such as digitizers and plotters were remotely connected at different locations, and were often not available for hands-on use. Microcomputer technology has radically changed this state of affairs. Microprocessors took most of the capability coming from the large computer and put it into the terminal itself. Over time more and more functions and capabilities have been moved from the mainframe computer to the graphics terminal. The next logical step was to disconnect the graphics terminal from the mainframe, a move to a self-contained graphics device called a *workstation*. A graphics workstation can support multiple input and output capabilities; may have its own RAM, own disk, and usually even its own operating system. To gain access to mass storage, and to communicate with other workstations, the workstation is usually networked.

Figure 2.07   A Typical Workstation (Photo Courtesy of GEMS, Cambridge, England)

A typical workstation for cartography (Figure 2.07) may include a digitizing tablet, a mouse and a keyboard for input, a color and text graphic screen for output, and could be networked to plotters, laser-jet and ink-jet printers, and color graphics cameras. The workstation concept is at the heart of the international graphics programming standard, GKS. In a workstation, any number of input and output devices can be supported as peripherals to the workstation. Graphics input devices can be locators, in which case they can point to locations on a map, or valuators, in which case they can simply indicate a value such as a dial or a joystick. For menus, input can be choice, in which case the user selects from a set of options, or pick, in which case the user can select by pointing to a segment of the drawing on the screen, such as a box or an object. Graphics output devices can literally be almost anything capable of producing a map. Using GKS, the concept of real and virtual maps finds expression in the terminology of the workstation, since a real map is simply a virtual map which has gone through the correct workstation transformations.

No discussion of hardware can be complete without a glance at the future and the new display technologies it holds. The one projection which will survive all the new technologies, however, is the workstation concept, which is versatile enough to incorporate any new hardware technology. Some devices do, however, appear to have a cartographic future. First, a color screen shutter is now available, which can be coupled with user-worn glasses to generate stereo screen images. In photogrammetry, remote sensing, and in terrain analysis, this seems to be a device with some interesting prospects, as yet unexploited. For example, with a stereo screen, hidden-line processing is unnecessary. Second, one vendor now is advance-testing a head-worn display which is small enough to be worn while driving, flying, or performing other tasks. The ability to generate maps in real-time so that they can be seen in peripheral vision is an exciting development and may rival the success of dashboard displays for vehicle navigation systems. Third, the linking of display technology with the real-time capabilities of satellite-based navigation systems has great potential for cartography. While none of these technologies will influence how maps are made, the potential for influencing map use is extraordinary.

2.3 REFERENCE

MOELLERING, H. (1980) "Strategies for real time cartography," *Cartographic Journal*, vol. 17, no. 1, pp. 12-15.

3

Software

3.1 SOFTWARE TRENDS

In Chapter 2, we examined the physical tools available for computer cartography. These tools, however, are merely expensive doorstops without the sets of computer instructions designed especially to allow the use of hardware for making maps. The computer instructions which interact with the cartographer, and allow the manipulation of data, hardware, and design for computer mapping, are collectively known as software. Computer cartographic software includes all of the data, programs, packages, and interactive systems necessary for mapping.

Several general themes pertain to software. First, as computers have moved from large mainframes alone to a diversity of mainframes, minicomputers, and microcomputers, computer mapping software has tended to become available over the full range, so that capabilities which were once available only on mainframe computers are now available to all. Computer mapping has moved away from mainframe computers toward smaller and more efficient microcomputers and workstations with dedicated graphics capabilities. Software has reflected this trend, with a movement away from large special purpose mapping programs to small, general purpose programs.

The early computer mapping software programs other than the large commercial systems tended to be single-purpose, stand-alone programs. As software has moved away from the large computer companies and academia, the programs have become more integrated, sharing a user interface, common data structures, and capabilities. Many automated mapping systems now are really just parts of much larger systems, such as statistical analysis or surveying systems. In addition, computer cartography now overlaps with the field of GIS. GISs are integrated systems for the management and analysis of spatial data with computer mapping capabilities. Many systems designed for image processing, for computer-assisted design, and for data presentation have computer mapping capabilities.

In 1974, the International Geographical Union's Commission on Geographical Data Sensing and Processing started to inventory the computer software for computer mapping, analytical cartography, and GIS (Brassel, 1977). Their three-volume report, published in 1980, was the first comprehensive survey of computer mapping software. Volume 1 was a catalog of complete geographic information systems. Volume 2 surveyed general geographic software and was entitled *data manipulation programs*. Volume 3, which surveyed cartography and graphics (Marble, 1980), was the result of a survey that was conducted during 1978-9, and as such gives a view of the types of computer mapping software available during the early years of computer mapping.

The report lists 38 complete mapping systems, defined as groups of programs with multiple mapping functions. In the list are several early prototypes of what later became GISs, which were initially developed as computer mapping systems. Twenty-five listings were systems for data collection, editing, and development. Editing systems were essential, because in the early days manual digitizing produced map data which were very error prone, and all translations, rotations, etc. had to be conducted by stand-alone secondary programs.

Twenty-seven systems performed basic drafting, operation, and design. These were systems which allowed you to draw in some interactive way on the computer screen. Many of these systems evolved into the basis for an entirely new industry by themselves, the computer-assisted drafting and design (CADD) industry, which overlaps into architecture, facilities mapping, and engineering. More recently, these systems have developed into microcomputer versions which put basic drafting capabilities, such as the manipulation of high-quality linework and text, into the hands of the microcomputer user. Many designers now use the so-called "paint" packages, and few drafting offices are without packages such as AUTOCAD, MacPaint, or MacDraw.

Nineteen of the packages displayed diagrams rather than maps, for example pie charts and histograms. While not specifically cartographic, these packages have become another new industry, called presentation graphics. The vendors in this area have made the presentation of statistical and other information into a communications art and have produced some very effective products which take advantage of hardware devices such as projectors and slidemakers.

Thirty-one of the systems did point and line mapping, producing maps of what is now called cartographic spaghetti. These systems were often dedicated to specific databases, such as the World Data Bank, or to a single type of plotter device. In 1980 many of the systems produced thematic maps, in particular the choropleth maps much in demand for statistical cartography. Here we see the split in technology, since 26 of the systems did area shading using grid or raster technology and 22 of them used vector technology. Almost all produced line printer output, printing and overprinting symbols to make patterns and shades.

Thirty-four of the systems did contouring and surface interpolation. Surface interpolation consisted of taking data collected as point samples with an attribute such as elevation or the depth of a drill hole and gridding the data for the purpose of making a contour map. Many of these systems have come about due to the needs of the oil exploration industry, yet produced very crude maps on a line printer. They used printed symbols, and lines had to be drawn in by hand afterwards. Few of them labeled the contours, rather, they printed legends along with the maps. The more sophisticated ones, packages like SURFACE II (Sampson, 1978), produced vector output, allowing smoothing and annotation, and had adequate user and reference documentation and support. Thirty-one systems were for three-dimensional mapping although most simply generated gridded perspective diagrams.

Sixty-seven systems did either map projection transformations, or other cartographic transforms, like translation, rotation, scaling, distance measurement, etc. These capabilities have found their way into most computer mapping software and rarely exist now as stand-alone programs. Ten of the systems produced cartograms, which are maps with geographic space distorted to show an attribute other than area, while seven of the systems were designed to do automated lettering. Seventeen of the "systems" were actually data; in other words, they were not software as such but files containing data. As a function, the supply of data has sometimes remained attached to software; for example, several computer mapping software vendors offer ready-to-use data for use with their systems.

3.1.1 Machine and Device Independence

One of the features listed in the IGU software survey was device dependency in the form of which computer the programs required. In 1980, the level of hardware dependence was substantial, with much of the software not only requiring particular word sizes, computers, programming languages, and compilers, but also specific input and output devices made by only one manufacturer. While in some cases this level of machine dependence remains today, far more items of computer mapping software run on a variety of microcomputers, minicomputers, and mainframes. Some run under multiple operating systems, a particularly important feature. Software packages also support long lists of different input and output devices, and even have installation menus and prompts that allow the support of multiple plotters, printers, digitizers, etc. Hardware vendors have realized the advantages of this, and often manufacture devices so that they emulate the protocols of the more popular devices.

Unfortunately, the absolute independence of machine and software is not yet complete. This means that as the market changes devices already purchased are sometimes left without applications software. When purchasing software, it is more important to match the software against mapping needs rather than against your existing hardware. Similarly, if hardware is available, the most general and highly supported devices are preferable.

3.1.2 Transportability

Machine and device independence, and the ability to move software and data between computers, are functions of the transportability of software. Transportable software can move between computers, operating systems, and devices with no more work than recompilation. Some computer programming languages are not transportable, with major variations from compiler to compiler and version to version. Software written in these languages will eventually disappear as programming environments change. Even operating system upgrades can make applications unusable, let alone the major changes that come about when new computers are installed.

The principal way in which transportability of cartographic software is ensured is by separating the graphics functions from the language. Even graphics functions change over time, and the only way to ensure against the loss of graphics capabilities is to use graphics standards, which will be supported over time. Within computer cartography, many vendors have been slow to write standards-based software, meaning that many computer mapping programs remain tied to a particular set of devices, operating system, or working environment. As highly portable languages such as C and Pascal form the basis of more mapping systems, and as graphics standards such as GKS, CORE, and PHIGS are used to build new software, users will be freer to transport software and even maps between computers and through operating systems.

3.2 TYPES OF SOFTWARE

3.2.1 Software for Large Systems

Computer mapping software for large computers is mostly that which runs on mini and mainframe systems. Among the many vendors are Synercom, Intergraph, and ESRI. Large computer systems often require special hardware, which can be purchased only through the software vendor. Many vendors manufacture their own workstations rather than depending upon other manufacturers, and supply software and hardware bundled together in "turnkey" or ready-made packages. Most minicomputers and mainframes have been used for graphics and mapping, but the more popular types are DEC's VAX and microVAX computers, and minicomputers by Prime, IBM, and others. These use a large variety of operating systems, such as DEC's VMS and Prime's PRIMOS. The lower end of the minicomputer market, however, has been completely overtaken by the workstation market. Leading workstation manufacturers are Apollo, Sun, Hewlett-Packard, and DEC. While the workstations use a variety of systems, most use the UNIX operating system. Languages common to many of these systems are Pascal, FORTRAN 77 (in which a large amount of computer mapping software is written), and increasingly, C. These languages and their compilers are becoming standardized over the various hardware environments due to the efforts of the standards organizations, especially the American National Standards Institute (ANSI). The use of an ANSI standard version of a language means that a program written on one computer and sent to another by tape, file transfer, or network link will compile without problems arising from the specifics of the machine on which the compiler is running.

3.2.2 Software for Small Systems

There are a great many computer mapping systems which run on microcomputers. It should be recognized, however, that microcomputers vary just as much as workstations and minicomputers in speed, size, capability, and peripherals. The first microcomputer mapping software appeared after the popularization of the microcomputer by Apple and then IBM during the very early years of the 1980s. In the few years since the origin of microcomputer software, the microcomputer has become infinitely more powerful. Many vendors have produced software to take advantage of this power. Like the industry itself, which has produced "shakedowns", some vendors have produced fine products and then disappeared completely. Among the first IBM-PC products was a choropleth mapping package called Atlas, by Strategic Locations Planning. This package allows the integration of polygon data for counties, census tracts, zip codes, and states, and supports choropleth mapping with a variety of shading patterns, colors, options for legends, etc. Atlas also supports map digitizing, so that data-bases other than those supplied by the vendor can be used. For Macintosh interactive mapping, the MacChoro package by Image Mapping Systems allows on-screen editing of choropleth maps, including some advanced text and legend design and interactive movement (Figure 3.01). Output for most microcomputer packages is to laser-jet printer or small flatbed plotter. A package with more varied mapping capabilities is the MapInfo package by the MapInfo Corporation. A long-surviving general purpose high-quality mapping package is GIMMS (Figure 3.02, p.37).

For terrain mapping, the SURFER package by Golden Software has some very sophisticated surface mapping capabilities and is highly interactive. This package includes interpolation of point data to a grid by distance weighting and by kriging, contour plotting, and three dimensional perspective grid generation (Figure 3.03 on p.38). These few microcomputer packages are the tip of a very large software iceberg. A list of microcomputer software suitable for the beginning and the advanced computer cartographer is included as Appendix A. This list is far from exhaustive. Information about new and existing software can be found as software reviews in the journals *The Professional Geographer*, published by the Association of American Geographers, and *The American Cartographer*, the journal of the American Cartographic Association of the American Congress on Surveying and Mapping. The prices of microcomputer software vary considerably, from tens to hundreds of dollars. Some large computer programs can cost in the thousands of dollars.

3.2.3 Related Software

The mapping software most closely related to computer cartography are the COGO or coordinate geometry packages. COGO packages were developed to automate the tasks done in surveyors' offices, which normally consisted of taking field measurements and producing plots. After collecting field data, many surveyors would bring their notebooks back to their offices to do all the angular calculations and conversions with a pocket

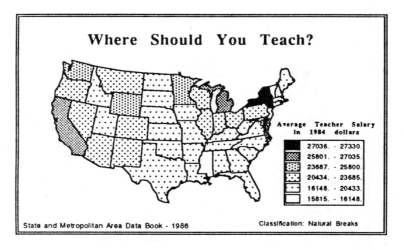

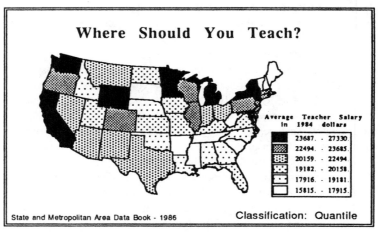

Figure 3.01   Sample output from the MacChoro Choropleth mapping software

calculator. The set of calculations would be passed on to a draftsman or cartographer, who would turn it into a map for use in planning, construction, or real estate. Some of the more sophisticated COGO packages do the field data reduction, perform the plotting, and can do cut-and-fill computations, produce legal property descriptions, etc.

Of the CADD systems, AUTOCAD by Autodesk has gained many cartographic followers, and has been extended to include many different mapping functions, for example as FMS/AC by Dennis Klein and Associates (Figure 3.04, p.39). AUTOCAD has some advanced data import and graphical editing capabilities, which make it a particularly good driver for digitizing maps. AUTOCAD's DXF internal file format can be read and written by several other mapping programs and GISs.

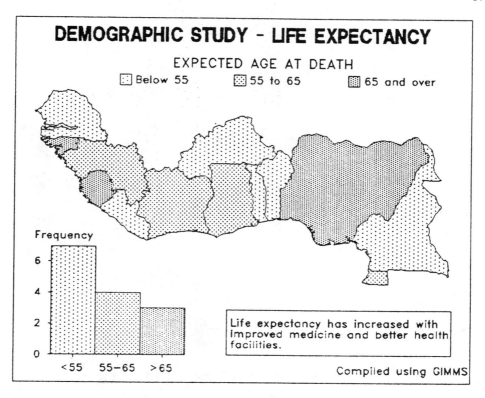

Figure 3.02   Sample output from the GIMMS computer mapping package (courtesy of Gimms Corp.)

The GIS software available normally has some advanced computer mapping capabilities. Among the higher quality in mapping terms is ARC/INFO by ESRI (Figure 3.05), which runs on minicomputers, workstations, and microcomputers. Some of the statistical packages also support mapping, among them SASGraph and SPSS Graphics.

Map projections still remains an area of interest to cartography. The University of Minnesota's WORLD package and the MICROCAM package by MicroConsultants, support interactive menus, multiple projections, and movable viewpoints, and drive various plotting devices (Figure 3.06, p.41).

3.2.4 Software for Software

Software development tools are pieces of software designed for writing computer mapping systems. They are designed to assist programmers, or in some cases advanced users, in producing their own software for automated mapping. Early software development tools were simply large collections of FORTRAN subroutines to support mapping and graphics. Among the suppliers of mainframe and microcomputer graphics toolboxes are Precision Visuals (DI-3000 and GKS-2000), UNIRAS, and Halo. Many of

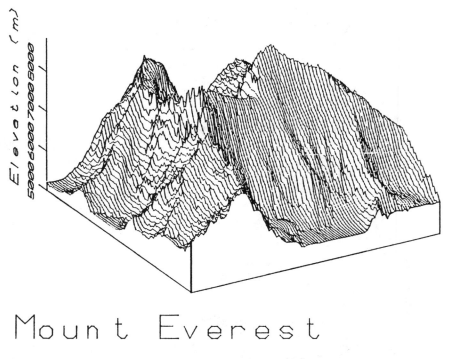

Figure 3.03   Sample output from the SURFER terrain mapping package

these packages are devoted to a particular device, and a particular graphics card, such as the EGA graphics on an IBM-PC. Support is usually by providing program-callable functions which plot points, lines, and areas, allow color selection, plot symbols, and text, etc.

In the chapters to come, a particular software tool, the GKS graphics standard, and the various bindings of this standard to the C language will be discussed. The strength of this tool is its device independent nature, i.e., the fact that the production of maps on a specific device is controlled by a separate piece of software called a device driver. The device driver interacts with more generic software, the standard, which does most of the map production. This system is highly flexible and is well suited to the demands of both computer and analytical cartography. Appendix B lists some of the vendors of GKS packages.

3.3 SOFTWARE RESOURCES

Computer cartography is a rapidly changing field. Anyone who purchases software can almost guarantee that next year will bring more and better alternatives. Computer mapping software is also expensive, although prices are continuing to fall. An alternative

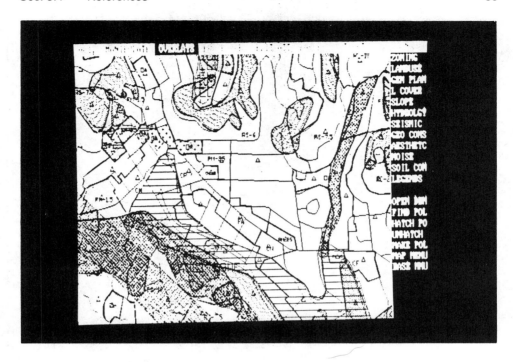

Figure 3.04   Sample map output from the FMS/AC AUTOCAD Facility Mapping
System (Photo Courtesy of FMS/AC)

source of software is the user group and shareware. Shareware is software supplied at
cost or at a nominal fee to anyone interested. In the United States a major clearing
house for shareware is PC-SIG (PC-SIG, 1987). Software also circulates free on many
computer bulletin boards, and is published in journals such as *Byte*. Within the field of
geography, the Geography Program Exchange of Michigan State University maintains a
software data-base. The Association of American Geographers' Microcomputer Speci-
alty Group also distributes shareware. Many data-bases and computer programs are
available from these sources, although their quality, scope, and languages vary considerably. These sources are excellent tools for getting a close look at the capabilities of
computer mapping systems before spending money on a more comprehensive mapping
system. Cartographers in particular, especially those capable of producing computer
cartographic software, should make an effort to support these organizations.

3.4 REFERENCES

BRASSEL, K. E. (1977) "A Survey of Cartographic Display Software," *International Yearbook of
Cartography*, vol. 17, pp. 60-76.

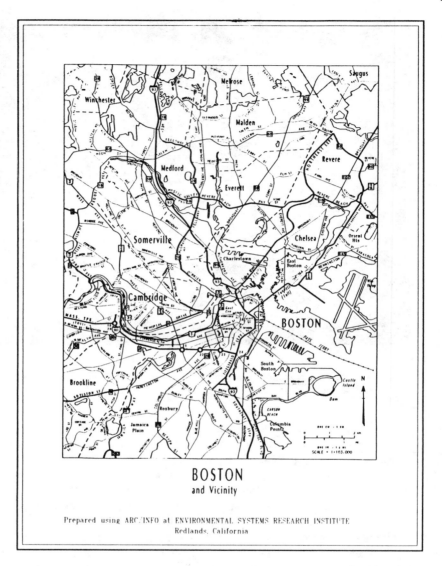

Figure 3.05   Sample output from ARC/INFO (used with permission of ESRI Corp.)

MARBLE, D. F. (Ed.) (1980) *Computer Software for Spatial Data Handling*, vol. 3: *Cartography and Graphics*, International Geographical Union, Commission on Geographical Data Sensing and Processing, Ottawa, Ontario.

PC-SIG (1987) *The PC-SIG Library*, PC-SIG Inc., Sunnyvale, CA.

SAMPSON, D. (1978) *The Surface II Graphics System*, Lawrence, KS, Kansas Geological Survey.

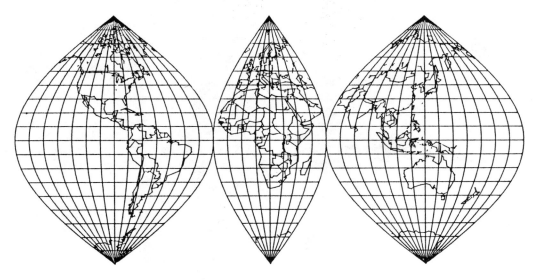

Interrupted Sinusoidal projection as used by the USGS~

Figure 3.06   Sample output from MICROCAM (courtesy of Microconsultants, Inc.)

4

Geocoding

4.1 GEOCODING AND COMPUTER CARTOGRAPHY

Geocoding is the conversion of spatial information into computer-readable form. As such, it is geocoding, both the process and the concepts involved, that determines the type, scale, accuracy, and precision of digital maps. An important aspect of geocoding is that effective geocoding requires an understanding of some basic geographic properties underlying geographic data. Unfortunately, it is quite easy to convert information into computer-readable form as data, but not all data are information, and all too frequently analytical cartography drowns in a sea of meaningless and unusable numbers. Most essential for analytical and computer cartography is that the information is about geographic phenomena. The role of the cartographer is to subdivide the broad landscape elements of geographic interest into smaller units more suitable for mapping, termed here *cartographic entities*. The cartographic entities can then be photographed, measured, sampled, or surveyed, and entered into computer mapping systems via *geocoding* methods to become *cartographic objects*.

Geocoding is the first stage of computer cartography and is done for two reasons. First, we can geocode with a specific mapping purpose in mind, for example capturing lines which are going to be parts of polygons to be used to produce a choropleth map of Western Europe. Second, we can do general purpose geocoding; in other words, we can collect every piece of information we can about a specific area, and assemble it for service in many possible cartographic contexts. The geocoding of general purpose cartographic data has great potential importance, just as accurate mapping of the country by topographic survey was important. The agencies performing this type of data collection and encoding have traditionally been government agencies, and we are fortunate that these agencies have often been at the forefront in making geocoding effective and efficient.

We can also geocode at two different levels. First, we can simply convert the graphic elements of a map into numbers so that we can reproduce the map using the methods of computer cartography. Many geographic data bases consist of geocoded data in this form. Alternatively, we can encode important topological information about the data we are geocoding. Increasingly, as GISs become the principal users of geocoded data, the encoding of topology becomes essential to the use and survival of cartographic data sets. To understand this distinction, a good example is the symbolization of roads and rivers. If the digital map consists of a road and a river which cross, a non-topological approach would be to plot the two lines one on top of the other. If the topology is encoded, we could recognize that either the road crosses the river on a bridge, in which case we draw a road symbol only and break the river symbol, perhaps adding the symbol for a bridge. Alternatively, the river could cross the road, implying an aqueduct or tunnel. The importance of topological geocoding is considered later. First, however, we will return to the idea mentioned above — that effective geocoding comes from a clear understanding of the fundamental geographic properties and their manifestations.

4.2 CHARACTERISTICS OF GEOGRAPHIC DATA

If the purpose of geocoding is to digitally encode the fundamental characteristics of geographic data, then we must understand what those characteristics are before we start designing strategies for capturing them. Fundamental to geographic data is the attribute of *location* on the earth's surface. While we have to use a third dimension to describe elevation on the earth, the two dimensions of location on the plane or sphere are probably the basic geographical property. Normally, x and y values represent latitudes and longitudes, but often in geocoding we assume a map projection such as a transverse Mercator, and we use coordinate systems such as Universal Transverse Mercator or State Plane to give locations. Occasionally, polar coordinates are used, giving an angle sometimes clockwise from north, and a distance. Most coordinate systems for use with computer mapping are based on cartesian coordinates. This implies that the axes of the two directions, such as eastings and northings, are orthogonal, that is, they are at right angles to each other. This allows us to specify a location in space by referring to a pair of coordinates (x,y), or an easting and a northing. Leaving out one or both of these means that a location is simply undefined. Location is therefore the most fundamental characteristic of both cartographic and geographic data.

A second fundamental characteristic of geographic data is *high volume*. Computer science traditionally has dealt with small data-bases by cartographic standards. Cartographic and geographic data-bases contain thousands or sometimes millions of data elements. A typical application in remote sensing uses seven one-byte bands of attributes and image arrays typically 512 by 512 in size. This represents a quarter of a million points for which we have multiple pieces of information, all of which may be required for a single display. Many cartographic data processing problems are generic problems of large data sets. As a result, in computer cartography we often have to deal with memory constraints and the efficiency of our data structure.

Fortunately, the cost of storing data has decreased dramatically. Even on small computers, new storage methods have increased available memory from kilo- to giga-bytes within only a decade. The effect has been to change the emphasis from simple storage volume to storage access time as the primary volume-related consideration, although memory constraints will remain a factor.

A third fundamental characteristic of geographic data is *dimensionality*. Tradi-tionally, cartography has divided data into *points, lines*, and *areas*. Somewhat related is the concept of level of measurement. Levels of measurement are divided into *nominal, ordinal, interval*, and *ratio*. These two divisions have formed the basis of at least two classifications of mapping methods and cartographic data, and will be considered in more detail in Chapter 8.

A fourth major characteristic of geographic data is *continuity*. Some map types, such as contours, assume a continuous distribution, while others, such as choropleth mapping, assume a discontinuous distribution. Continuity is an important geographical property. The best example of a continuous variable is probably surface elevation. As we walk around on the earth's surface, we always have an elevation. There is no point where elevation is undetermined. In real terrain, there are very few exceptions. Vertical overhangs and cliffs do indeed have areas where elevation is locally undetermined, but on the whole, elevation as a geographic distribution is continuous. Continuity does not always apply to statistical distributions. For example, tax rates are a discontinuous geo-graphic variable. A resident of New York has to pay the state personal income tax, but by living just 1 meter inside Connecticut, there is none. The tax rate is a discontinuous geographic variable because on the boundary line, the tax rate is undefined. It was once somewhat facetiously suggested that the only truly discontinuous geographic variables were tax rates and road surfaces, as anyone who has paid New York taxes or driven into the city of New York knows.

In addition, geographic continuity is an important property. Space classifications by areas must be exhaustive for continuity, i.e., there should be no holes or unclassified areas. Similarly, for a set of categorical attributes reflecting a map, the set should con-tain all the objects found on the map, without any "other" or "miscellaneous" categories.

4.3 FUNDAMENTAL PROPERTIES OF GEOGRAPHIC DATA

So far we have discussed the characteristics of geographic data. We should now con-sider some of the fundamental underlying properties. While characteristics of geo-graphic data influence computer cartography, the implications of the fundamental pro-perties are closer to the concerns of analytical cartography.

A basic underlying property is *size* and its characterization in measurement. Most geographic phenomena can be measured directly, for example by survey or air photo. A point has the measured aspects of location (x,y), adjacency, and elevation. A line has length, direction, connectivity, and "wigglyness". A polygon has topology (whether there are holes or outliers), area, shape, boundary length, as well as location and orientation. A volume has topology, continuity, surface slope, surface aspect, surface trend, structure, location, and elevation. Most of these properties are comparatively simple to measure if the cartographic data are geocoded. Some are extremely difficult to measure in the real world, and as such can only be analyzed using the data abstractions of mapping. These measurements can usually be implemented by simple algorithms, some of which are given in the following chapters.

Another fundamental property is *distribution*. Density is a measure of the distribution of a phenomenon across space. Density can be computed by counting cartographic objects or their attributes over a set of geographic units, such as a grid or a set of regions. The density of a geographic phenomenon has a great many implications not only for how we measure and geocode it, but also how we can generalize it, decide on map coverage, and symbolize it on a map.

Another fundamental property of geographic data is *pattern*. Pattern is actually a characteristic of distributions and is a description of their structure. Pattern can be thought of as a lack of randomness. A "first law of geography" is that anything that is of geographic interest lies at the intersection between two and four maps, photos, or images. This relates to distribution via the sampling theorem. The "second law of geography", with apologies to Tobler, is that everything is related to everything else, but *near* things are more related than others. In other words, if two things are close to each other, they are more likely to be similar than if they are separated by a long distance. The simplest way in which this "nearness" phenomenon can be seen and measured is by repetition. Those relationships that repeat themselves at distances of less than half the size of the map result in patterns, implying that repetition is very important for pattern. Lack of randomness can actually be measured. For simple point distributions we can actually measure randomness, or lack of randomness, by using the nearest-neighbor statistic. A test for pattern goes beyond simple measurement, however, since to observe pattern we need a model or description of the pattern we wish to find. In remote sensing and image processing, we can pass "templates" over a continuous distribution looking for a match of the discontinuous feature we wish to detect. This approach works even if the image is obscured by error or atmospheric effects. Patterns repeated through scales are also of interest, and can be called self-similar.

The *neighborhood property* is, along with pattern, one of the most definitive of geographic properties. If pattern is the repetition of an attribute over space, the neighborhood property defines how the property varies over space. A key aspect of the neighborhood property is that variation takes place with distance, so little separations mean similarity, and big separations mean dissimilarity. Geographic studies often examine the relationship between some geographic phenomenon and distance. Usually we see the neighborhood property as a distance function, a mathematical expression of the relationship between a geographic property and distance. Within geography, this distance function has been characterized and measured using tools such as the autocorrelation function, spatial interaction models, distance decay models, and the variogram. These functions express mathematically exactly what the second law of geography states.

The variogram is the simplest expression of the distance relationship. Imagine a map showing three elevation benchmarks. Geometrically, these three elevations define a plane. Geographically, however, a plane is not a realistic distribution. At any place on the map, the elevation is related somehow to the three benchmark elevations, and according to the neighborhood property, to the distances from them. Starting at any given point we could determine the difference in elevation between this and all other points. Since some elevation differences are positive and some negative, we could simply square the values to get an absolute number, and sum over the set of points. One half of this number is called the semi-variance, and a plot of semi-variance against distance is called a semi-variogram. The semi-variogram gives us an expected value for an elevation deviation at any given separation between points. Some semi-variograms are shown in Figure 4.01.

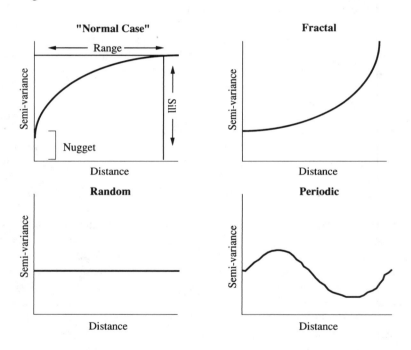

Figure 4.01   Some semi-variograms

Some important features are the nugget variance, the sill, and the range. The nugget variance is the semi-variance we can expect at zero distance. This can be interpreted as the measurement error associated with our method for determining elevation, which is not always zero. The form of the semi-variogram is determined by the neighborhood property, since the relationship between the influence of neighboring objects upon each other is measured directly.

For example, imagine starting at one of the benchmarks with a known elevation. Next, we could take a walk in a straight line off in a randomly chosen direction. How is the elevation where we end up related to the elevation where we began? The neighborhood property suggests that the farther we go, the less our elevation is determined by the elevation where we started. For a very long trip, say across several states, the elevation where we end up has almost nothing to do with where we started. If we just take a short step, we are not going to change elevation much. This does assume continuity of elevation. We could have taken a step off a cliff, but the chances are we will not change elevation very much over short distances.

This implies that the form of a typical variogram is a line increasing to the right, that is, that semi-variance increases with distance. However, there is only a finite range of elevations on the earth's surface. As the distances get very high, there is no increase in semi-variance. The distance where the semi-variogram flattens out is called the range, and the value of the semi-variance where this takes place is the sill.

Different geographic phenomena have different semi-variograms. If we never reach the sill, we are looking at the wrong map scale, and our measurements do not cover a big enough map area to see the phenomenon we are looking at. A perfectly flat plane would have no elevation variation, so the semi-variance at all points would be zero. More realistically, the nugget variance would extend over the entire range and the semi-variogram would be flat, or "silly". The same would be true of a surface showing only random variations in elevations. This type of variogram reveals a lack of spatial form and no underlying geography. The analyst should either look elsewhere, change the map scale, or give up. This is not a geographic phenomenon. To be geographic, it has to be non-random; it has to have some pattern, some expression of the neighborhood property; otherwise, it is of no geographic interest.

Occasionally, a semi-variogram reveals a periodic function. A periodic phenomenon is something that repeats itself, showing spatial pattern. Imagine elevations in Pennsylvania. If we pick any random point and start walking, we are going to start to have a normal-looking variogram. Then suddenly, when we reach some key distance, our elevations become similar to our starting elevations. This is because we are in a ridge and valley topography and elevations form periodic cycles. The elevation differences squared computed for the semi-variogram again approach zero, particularly across the fold of the terrain. A further example would be elevations over a dune field, except that with dunes the directionality would be less. Also characteristic is the curve as it climbs toward the sill — its form, gradient, and relationship to direction.

Contiguity is another important geographic property. Contiguity is the property of being related by juxtaposition. Contiguity, therefore, is one of the geographic expressions of topology. The best example of contiguity is the sharing of a common boundary. Political geographers may be interested in the length of the border of Poland, perhaps even the geometric shape of Poland, but the main item of geographic interest would be a list of nations with a common boundary. Similarly, in land cover terms, we may be interested in which land uses are most likely to be found around lakes. Whether the answer is beach cottages or swamps, we have a distribution of direct geographic interest.

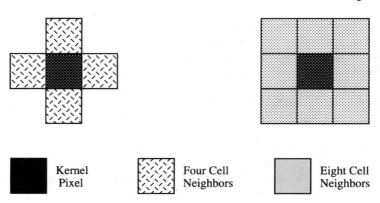

Figure 4.02   Contiguity of pixels

Contiguity is expressed in many ways. We can define it in terms of shared boundaries, in which case we can measure the number and lengths of shared boundaries. Within networks, contiguity is referred to as connectivity. For example, a network may contain connected links between nodes. In the sense of the network, the nodes are contiguous, i.e., spatially you can get there from here in one step. Also, often we think of contiguity in terms of pixels, or a grid structure. In fact, we do this so frequently that we even have special terms for it: we talk about four-cell and eight-cell contiguity (Figure 4.02). The center pixel for a neighborhood is called the kernel and the contiguous pixels are those which share a direct common boundary.

Shape is another very geographic property. Shape is a difficult property to measure directly, so most shape measures are really measures of the level of correspondence between shapes. Some, however, are graphic with fewer dimensions. An excellent review of shape measurement is that by Pavlidis (1978). Scientists have measured the shapes of many phenomena. For example, geologists have measured the shape of Pacific atolls in the ocean, and biologists have measured the shape of cells and the wings of butterflies. Many geographers are interested in the shape of cartographic objects, such as the shapes of congressional districts or geomorphological features.

It is best to illustrate the property of shape by example. A particularly simple shape measure is that of Lee and Sallee (1970), illustrated in Figure 4.03. This measure chooses a point somewhere inside the shape and draws a circle with the same area as the shape. The two figures are then overlain, forming three types of regions, an overlap or intersection, and then remaining parts of each of the two shapes. In set theory, the shape and the circle can be termed A and B. Then the Lee and Sallee shape measure is expressed as

$$s = 1 - \frac{A \cap B}{A \cup B} \qquad\qquad 4.01$$

Notice that for a circle, A union B and A intersection B are both equal to 1. Subtracting 1 over 1 from 1 gives zero. So the base measurement is zero, and the shape measure compares shapes to a circle. Note, however, that the shape number depends on the

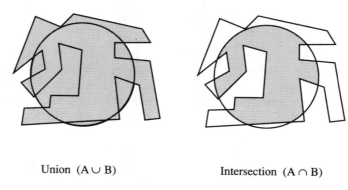

Union (A ∪ B)   Intersection (A ∩ B)

Figure 4.03   The Lee and Sallee shape measure

area of the chosen "comparison shape", the point at which the two figures are overlain, and the orientation. The same measure could be used to compute the resemblance between any set of shapes, such as the resemblance of the remaining 49 state boundaries to the shape of New York. There are many other shape measures. Shape, however, is a very complex property. For example, the measure described above could give the same number to an infinite number of different shapes. An inversion of the method, for example to perform recognition, would be impossible.

The final fundamental geographic property is *scale*. As far as cartography is concerned, the property of scale is the most distinctive feature of things that are geographic. The simplest expression of scale is the representative fraction, the ratio of distances on a map, to the same distances on the part of the world shown on the map. There is a limited range of scales over which cartography has an interest in a phenomenon. If instead of thinking about things at a particular scale, we think about things through scales, one of two things can happen. Either objects become more clear at certain scales, or they never change with scale. These two sub-properties are called scale dependence and scale independence.

If things do not change with scale, they can be modeled using fractal geometry. Fractal geometry has dimensions that are not whole numbers. If we go back to our original geographic characteristic of dimension, we can consider the point, line, area, and volume as special geometric cases. In fractal geometry, lines have dimensions between 1.0 and 2.0, and areas and surfaces have volumes between 2.0 and 3.0. Scale dependence means that there is an appropriate scale for making a map. Analytical cartography can contribute by determining this scale objectively.

4.4 GOALS OF GEOCODING METHODS

Geocoding seeks many conflicting goals. Each should be kept in mind before geocoding cartographic data, and the computer cartographer should be realistic about which goals are at odds with each other.

4.4.1 Minimize Labor Input

Since one of the major sources of error in the geocoding process is human error, and since labor costs are usually high, especially for semi-automatic digitizing, an important goal for geocoding is to minimize the amount of manual labor involved. The manual component for converting existing paper maps is high and can multiply unexpectedly. Chrisman (1987) reported manual digitizing times of eight hours per 300 polygon soil map sheet, with an additional four hours of editing, even using software designed explicitly for the reduction and detection of digitizing errors, with automatic topological correction and automatic end-node snapping. Chrisman pointed out that careful attention to the editing capabilities of digitizing software, coupled with consistency checking, can allow even a low-cost microcomputer workstation to produce high-quality, accurate digital cartographic data.

For example, we may seek a digital land cover map at 1:500,000. One approach would be to draw a grid with squares of 20 mm, representing 5 km on the final grid after scaling over four 1:250,000 land use and land cover maps from the United States Geological Survey. We could then go grid cell by grid cell writing down the land use category at each intersection. We could then type the land use category numbers one by one into a computer file. Assuming that we can draw the grids and write down the numbers at the rate of 200 per hour, and can enter the data at the same rate, we would need about 400 hours, or 10 weeks of full-time labor, to geocode the map, without any checking. While in actuality, geocoders become faster at a task with experience, anything we can do that makes data entry easier, we should do. In many cases, simple steps can be taken. Digitizing tablets that are adequately lit, and are repositionable, especially so that the user can sit down, can save substantial amounts of labor time, since much work is necessary going back over errors and cross checking. Immediate video and audible feedback for errors can save starting over after a string of errors, as also can direct editing, off-line control, and a quiet, non-distracting workplace.

More substantial steps can also be taken to reduce labor. Many steps are totally unnecessary, such as the writing down on paper in the example above. If in doubt, the best way to estimate the time for each step of the digitizing process is to conduct a small test on a pilot data set, taking it through the entire procedure. The time can then be multiplied up to get overall estimates.

4.4.2 Detect and Eliminate Errors

Many early geocoding systems had only limited editing capabilities. They allowed data entry, but error detection was by batch processing and correction was by deletion and reentry. Anything we can do in the geocoding process that reduces errors, or that makes errors easily detectable, we should indeed do. As an absolute minimum, data for lines and areas can be processed automatically for topological consistency, and any unconnected lines or unclosed polygons detected and signaled to the user.

The easiest way to avoid errors in geocoding is to make sure that they are detected as soon as possible, and then to make their correction easy. Video display during digitizing, and audio feedback for error messages are essential. Software should spell out exactly what will happen in the case of an error. A common geocoding error is to overflow a hard or floppy disk while digitizing. Some software continues to accept data as if nothing is wrong, and gives a "disk full" message only when you exit from the digitizer software.

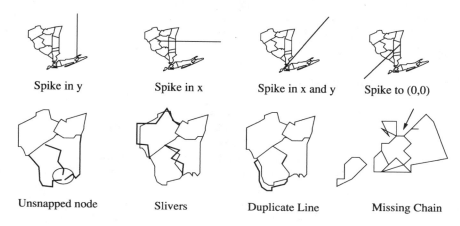

| Spike in y | Spike in x | Spike in x and y | Spike to (0,0) |

| Unsnapped node | Slivers | Duplicate Line | Missing Chain |

Figure 4.04   Some easy-to-detect geocoding errors

Some easy-to-detect errors are slivers, spikes, inversions, and disappearing nodes (Figure 4.04). Others take more effort. Some rules of thumb for the most difficult errors are:

(1) If something looks wrong on the display, it is probably wrong in the data.

(2) Believe the locational data files before the display.

(3) Believe the source map before the data.

(4) Stopping or having another person work on geocoding often is a source of errors.

(5) Postponing a correction gives the correction only a 50% chance of ever being made unless post-processing for errors is available.

(6) Sloppy geocoding makes unreliable and erroneous maps.

In the section of the digital cartographic data standards on data quality, a distinction is made between positional and attribute accuracy and logical consistency. Positional and attribute accuracy can be tested and measured by direct comparison with the source. Very useful aids in detecting errors are overlay plots at the same size as the source map, especially if they can be superimposed on the original. There is no substitute, however, for systematically parsing through the geocoded data looking for discrepancies. Often, plotting the data becomes a useful aid since unplottable data often have bad geocodes. Similarly, attempting to fill polygons with color often detects gaps and slivers not visible in busy polygon networks. The best check for positional accuracy is a check against an independent source map of higher accuracy.

A data set which is correctly geocoded both positionally and with attributes is not necessarily logically consistent. Logical consistency can be checked most easily for topological data. Topologically, data can be checked to see that all chains intersect at nodes, that chains cycle correctly in a ring around polygons, and that inner rings are fully enclosed within their surrounding polygons. Otherwise, attributes can be checked to ensure that they fall within the correct range and that no feature has become too small to be represented accurately.

4.4.3 Optimize Storage Efficiency

Data contains two parts: volume and information. Geocoding often seeks to eliminate volume while retaining information. We need to know what parts of the data are redundant, what can be left out, and what must be kept.

The ideal storage method for geocoded data saves only the minimum, the distilled information. However, redundant information is often a necessary part of geocoding. A state boundary on one map projection may become curved on another, and would need many points to make it look curved, many more than the "optimal" two necessary for a straight line. In addition, storage efficiency depends entirely upon which data structure we will be using to map the digital data, or how we will move the raw geocoded information between systems or applications.

4.4.4 Maximize Flexibility

A critical goal is to optimize flexibility. The most elegant data structure in the world is worthless if the author is the only one who can read it, or wants to read it because it is so obscure. We need data structures and storage methods which allow analyses that we did not anticipate when we geocoded the data originally. These unanticipated data uses are not the exception, they are the norm. As soon as data exists in digital form, new uses for the data will be discovered, each of which will place a new set of demands on the chosen data structure. A measure of the flexibility of a data structure is how well it holds up to these unanticipated demands, rather than the original uses for which it was designed.

4.5 LOCATIONAL GEOCODES

Geocoding uses a map coordinate system. Coordinate systems can be standard or arbitrary. There are many reasons to avoid arbitrary referencing systems, among them incompatibility, lack of ability to document the alternate systems, inability to adapt to new levels of precision, or the assumption of a specific map projection. An important factor in choosing a coordinate system with which to geocode a map is its universality. A good coordinate system works world-wide, is simple, accurate, precise, terse, and adaptable. Unfortunately, the problems of global coordinate systems are much like those of map projections; that is, not all goals can be served at once. As a result, we have several established systems for specifying locations on the earth's surface.

4.5.1 The UTM Coordinate System

If we use the ability to georeference other planets as a measure of success, a successful geocoding system for cartography is the Universal Transverse Mercator (UTM) coordinate system. The equatorial Mercator projection, which distorts areas so much at the poles, nevertheless produces minimal distortion along the equator. Lambert modified the Mercator projection into its transverse form in 1772, in which the "equator" instead runs north-south. The effect is to minimize distortion in a narrow strip running from pole to pole. UTM capitalizes on this fact by dividing the earth up into pole-to-pole strips or zones, each 6 degrees of longitude wide, running from pole to pole. The first zone starts at 180 degrees west (or east), at the international date line and runs east, that is from 180 degrees west to 174 degrees west. The final zone, zone 60, starts at 174 degrees east and extends east to the date line. The zones therefore increase in number from west to east. For the United States, California falls into zones 10 and 11, while Maine falls into zone 19.

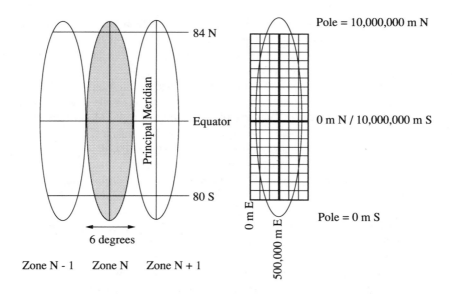

Figure 4.05   The Universal Transverse Mercator Coordinate System

Within each zone we draw a transverse Mercator projection centered on the middle of the zone. Thus for zone 1, with longitudes ranging from 180 degrees west to 174 degrees west, the central meridian for the transverse Mercator projection is 177 degrees west. Since the equator meets the central meridian of the system at right angles, we use this point to orient the grid system (Figure 4.05). Two forms of the UTM system are in common use. The first, used for civilian applications, sets up a single grid for each zone. To establish an origin for the zone, we work separately for the two hemispheres. For the southern hemisphere, the zero northing is the South Pole, and we give northings in meters north of this reference point. Fortunately, the meter was originally defined as one ten millionth of the distance from the pole to the equator, actually measured on the meridian passing through Paris. While the distance varies according to which meridian is measured, the value 10 million is sufficient for most cartographic applications. Although the meter has been redefined in a more precise way, the student may wish to compare the utility of its origin with the origin of the foot, first standardized as one third of the distance from King Henry the First's nose to the tip of his fingers.

The numbering of northings start again at the equator, which is either 10,000,000 meters north in southern hemisphere coordinates or 0 meters north in northern hemisphere coordinates. Northings then increase to 10,000,000 at the North Pole. Note that as we approach the poles the distortions of the latitude-longitude grid drift farther and farther from the UTM grid. It is customary, therefore, to use the UTM system neither beyond the land limits of North America, nor for the continent of Antarctica. This means that the limits are 84 degrees north and 80 degrees south. For the polar regions, the Universal Polar Stereographic coordinate system is used.

For eastings a false origin is established beyond the westerly limit of each zone. The actual distance is about half a degree, but the numbering is chosen so that the central meridian has an easting of 500,000 meters. This has the dual advantage of allowing overlap between zones for mapping purposes, and of giving all eastings positive numbers. Also, we can tell from our easting if we are east or west of the central meridian, and therefore the relationship between true north and grid north at any point. To give a specific example, Hunter College is located at UTM coordinate 4,513,410 meters north; 587,310 meters east; zone 18, northern hemisphere. This tells us that we are about four tenths of the way up from the equator to the pole, and are east of the central meridian for our zone, which is centered on 75 degrees west of Greenwich. On a map showing Hunter College, UTM grid north would therefore appear to be east of true north.

For geocoding with the UTM system 16 digits is enough to store the location to a precision of 1 meter, with one digit restricted to a binary (northern or southern hemisphere), and the first digit of the zone restricted to 0 to 6 (60 is the largest zone number).

This coordinate system has two real cartographic advantages. First, geometric computations can be performed on geographic data as if they were located not on the surface of a sphere but on a plane. Over small distances, the errors in doing so are minimal, although it should be noted that area computations over large regions are especially cartographically dangerous. Distances and bearings can similarly be computed over small areas. The second advantage is that the level of precision can be adapted to the application. For many purposes, especially at small scales, the last UTM digit can be dropped, decreasing the resolution to 10 meters. This strategy is often used at scales of 1:250,000 and smaller. Similarly, sub-meter resolution can be added simply by using decimals in the eastings and northings. In practice, few applications except for precision surveying and geodesy need precision of less than 1 meter, although it is often used to prevent computer rounding error.

4.5.2 The Military Grid Coordinate System

The second form of the UTM coordinate system is the military grid, adopted for use by the United States Army (Department of the Army, 1967) and many other organizations. The military grid uses a lettering system to reduce the number of digits needed to isolate a location, since letters can be spelled out using simple words over radio broadcasts. Zones are numbered as before, from 1 to 60 west to east. Within zones, however, 8-degree strips of latitude are lettered from C (80 to 72 degrees south) to X (72 to 84 degrees north: an extended-width strip). The letter designations A, B, Y, and Z are reserved for UPS designations. A single rectangle, 6 by 8 degrees, generally falls within about a 1,000 kilometer square on the ground. These grids are referenced by numbers and letters, for example, Hunter College falls into grid cell 18T (Figure 4.06).

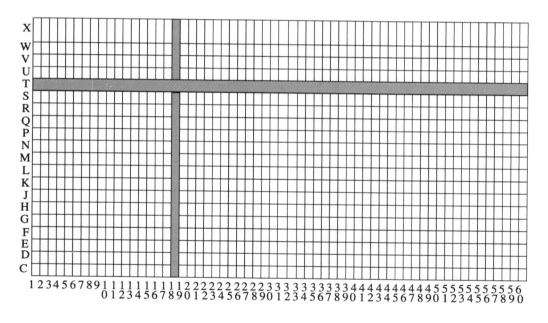

Figure 4.06   Six-by-eight degree cells on the UTM military grid

Each grid cell is then further subdivided into squares 100,000 meters on a side. Each cell is assigned two additional letter identifiers (Figure 4.07). In the east-west (x) direction, the 100,000-meter squares are lettered starting with A, up to Z, and then repeating around the world, with the exception that the letters I and O are excluded, since they could be confused with numbers. Thus the first column, A, is 100,000 meters wide, and starts at 180 degrees west. The alphabet recycles about every 18 degrees, and includes about six full-width columns per UTM zone. Several partial columns are given designations nevertheless, so that overlap is possible, and some disappear as the poles are approached.

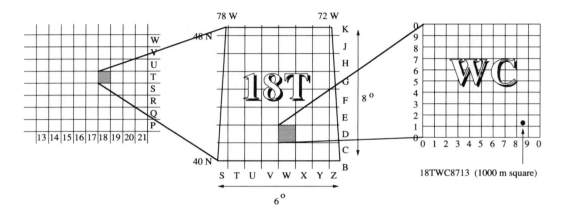

Figure 4.07   Military grid cell letters

In the north-south (y) direction, the letters A through V are used (again omitting I and O), starting at the equator and increasing north, and again cycling through the letters as needed. The reverse sequence, starting at V and cycling backwards to A, then back to V, etc., is used for the southern hemisphere. Thus a single 100,000-meter grid square can be isolated using a sequence such as 18TWC. Within this area, successively accurate locations can be given by more and more pairs of x and y digits. For example, 18TWC 81 isolates a 10,000-meter square, 18TWC 8713 a 1,000-meter square, and 18TWC 873134 a 100-meter square. These numbers are frequently stored without the global cell designation, especially for small countries or limited areas of interest. Thus WC873134, two letters and six numbers, would give a location to within 100-meter ground accuracy.

The universal polar stereographic coordinate system (UPS), also part of the military grid, is based on a stereographic map projection centered on each of the poles. The two projections are centered so that the western hemisphere is to the left. For the north polar region, the western zone is designated Y, and the eastern Z, each extending from 84 to 90 degrees north. The prime meridian is used as the right angle for the grid, and 100,000-meter grid cells are simply lettered from A to P running from the bottom (the prime meridian) to the top. The eastings for the letter designations Y and Z are chosen so that the first column left of the pole is Z and the first to the right is A. For cell Y, R is the first column, while for cell Z, J is the last. For the South Pole, the situation is identical, except that the cells are inverted, i.e., the Greenwich meridian is to the top,

are designated A to the west and B to the east, and since the zone is larger (80 to 90 degrees south), the letters go from A to Z as northings and J to Z and A to R as eastings, respectively (Figure 4.08).

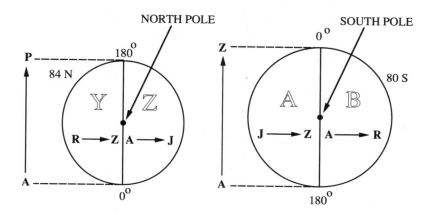

Figure 4.08   100,000-meter cells on the UPS grid

4.5.3 Geographic Coordinates

Many global data-bases record locations using latitude and longitude or geographic coordinates. Latitude and longitude are almost always geocoded in one of two ways. Latitudes go from 90 degrees south (-90) to 90 degrees north (+90). Precision below a degree is geocoded as minutes and seconds, and decimals of seconds, in one of two formats: either plus or minus DD.MMSS.XX, where DD are degrees, MM are minutes, and SS.XX are decimal seconds; or alternatively, as DD.XXXX, or decimal degrees. Longitudes are the same, with the exception that the range is -180 to +180 degrees. In the second format, degrees are converted to radians and stored as floating point numbers with decimal places.

It is especially important that we record how each map has been geocoded if geographic coordinates are used. Maps look particularly strange when decimal degrees are taken for the degree-minute-second format. The relatively open exchange of geocoded data makes this recording of the system even more important. The simple rectangular projection, in which latitude and longitude are simply drawn without projection at all, may get the map done, but denies a cartographic tradition going back over 1,000 years and may give incorrect results if used for computations.

4.5.4 The State Plane Coordinate System

Much georeferencing in the United States uses a system called the State Plane Coordinate System (SPCS). The SPCS is based on feet, and has been used for decades to write legal descriptions of properties and engineering projects. Legal documents are probably the least modifiable descriptions on earth. The SPCS is based upon a different map of each state, except Alaska. States that are elongated north to south, such as California, are drawn on a transverse Mercator projection. States that are elongated east to west, such as New York, are drawn on a Lambert conformal conic projection. The state is then divided up into zones, the number of which varies from small states, such as Rhode Island with one, to as many as five for large states. Some zones have no apparent logic; for example, the state of California has one zone that consists of Los Angeles County alone. Some have more logic, so, for example, Long Island has its own zone for the state of New York. Since there are so many projections to cover the land area, generally the distortion attributable to the map projection is very small, much less than in UTM, where it can approach 1 part in 2,000.

Each zone then has an arbitrarily determined origin which is usually some given number of feet west and south of the southwesternmost point on the map. This again means that the eastings and northings all come out as positive numbers. The system then simply gives eastings and northings in feet, often ending up with millions of feet, with no rounding up to miles. The system is slightly more precise than UTM, because coordinates are to within a foot rather than a meter, and can be more accurate over small areas. A disadvantage is the lack of universality. Imagine surveying the boundary between not only two zones, but two states. This means that you could be surveying an area that falls into four coordinate systems. Calculating areas on that basis becomes a set of special case solutions. On the other hand, SPCS is used universally by surveyors all over the country.

4.5.5 Other Systems

There are, in addition, many other georeferencing systems. Most countries have their own, although many use UTM or the military grid. The National Grid of the United Kingdom uses the lettering system of the Military Grid. In a few cases, particularly Sweden, the national census and other data are directly tied into the coordinates. Within the United States, many private companies and public services use unique systems, usually tied to specific functions such as power lines, or a specific region such as a city municipality. When using a georeferencing system for geocoding, we should be sure to remain consistent within that system and to record the relationship between the system and latitude and longitude, or some other recognized system. Also, we should be sure to use precision and numbers of significant figures which make sense. Can we really measure distances over entire states down to the micrometer or less? And even if we can, is this storage efficient? On the other hand, there is also a tendency to throw away precision needlessly. If we round all cartographic data up to 78 meters, a convenient number determined by an arbitrary orbiting satellite system, how will we deal with newer, higher-resolution systems?

In summary, the universality of the UTM system makes it attractive for world-scale geocoding, while for small areas other systems may be more accurate. Care should be taken to be consistent, accurate, and to use an appropriate level of precision.

Fortunately, computer software for mapping allows data to be used from more than one referencing system.

4.6 GEOCODING METHODS

Historically, many different means have been used to geocode. At first, some computer cartographic software actually required maps to be encoded and entered by hand. The hours of monotonous work required for this task makes errors common and their correction difficult. When special purpose digitizing hardware became available, and especially since the cost of this hardware fell substantially, virtually all geocoding was performed by computer.

There are two distinct technologies for geocoding, the first of which assists a person in performing the digitizing task, the second of which performs the task in a completely automated fashion. The first type, *semi-automated digitizing* involves the use of a digitizer or digitizing tablet (Figure 4.09). This technology has developed as computer mapping and computer-aided design have grown and placed new demands on computer hardware.

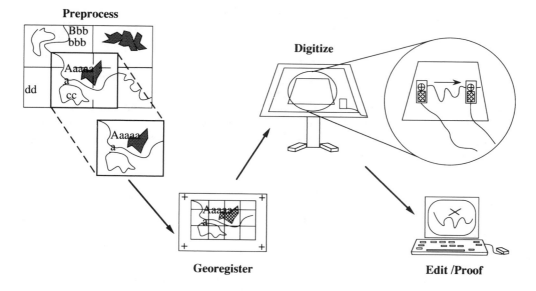

Figure 4.09   The semi-automatic digitizing process

The digitizing tablet is a digital and electronic equivalent of the drafting table. The major components are a flat surface, onto which can be placed or usually taped a map, and a stylus or cursor, which has the capability of signaling to a computer that a point has been selected. As we saw in Chapter 2, the mechanism to capture the location of the point differs. Many systems have connected arms, but most have embedded active wires in the tablet surface which receive an electrical impulse sent by a coil in the cursor. In some rare cases, the cursor transmits a sound, which is picked up and recorded by an array of microphones.

The actual process of digitizing a map proceeds as follows. First, the paper map is tailored or preprocessed. If the map is multiple sheets, then the separate sheets should be digitized independently and digitally merged (zipped) later. The next major step, unless annotations have to be made onto the map to assist the geocoding, is to derive a coordinate system for the map. Most applications use UTM, the Military Grid, or latitude and longitude, but many cartographers ignore these standard systems and use hardware coordinates or map inches or millimeters. As a minimum, the coordinate locations of two points are required, usually the upper right easting and northing and the lower left easting and northing. From these two points, with their given (user or world coordinates), and their raw digitizer coordinates, all of the parameters can be computed for affine transformations. This means that the orientation of the map on the tablet need not be exact as is required of some digitizing packages. Many software packages require three or even four of these control points for computing the affine transformation parameters (see Chapter 9).

The beginning of the digitizing sequence involves selecting the control points and interactively entering their world coordinates. This is a very important step, since an error at this stage would lead to a complex systematic error in every pair of coordinates. After the map is taped to the tablet, it should not be moved without reregistration, and it is preferable to perform the registration only once per map. Ideally, the entire digitizing process should be finished at one sitting, although this is often impossible. Tape should be placed at each map corner after smoothing the map, and care should be taken to deal with folds and the crinkles that develop during periods of high humidity with certain papers. A stable base product such as Mylar is preferable for digitizing. The lower edge, which will have the cursor and your right sleeve (if you are right handed) dragged over it many times, should be taped over its entire length. Always permanently record the x and y of the map extremes.

Digitizing then proceeds with the selection of points. The cursor may have multiple buttons, and may be capable of entering text and data without using the keyboard. Voice data entry and commands are also sometimes used. On specialized workstations, there may even be a second tablet with its own mouse or cursor for commands. Errors can be reduced during this process by reading the documentation in advance, and by frequently stopping to check the actual data being generated. Points are usually entered one at a time, with a pause after each to enter attributes such as labels or elevations. Lines are entered as strings of points, and must be terminated with an end-of-chain signal to determine which point forms the node at the end of the chain. This signal must come from the cursor in some way, either by digitizing a point on a preset menu-area or by hitting a preset key. Unless the chains are to be software-processed for topology, chain linkage information may need to be entered also. Polygons are usually digitized as chains, although sometimes an automatic closure for the last point (snapping) can be performed. Finally, the points should be checked and edited. The digitizing software may contain editing features, such as delete and add a chain, or move and snap a node. The software may also support multiple collection modes. *Point mode* simply digitizes one point each time the button on the cursor is pressed. *Stream modes* generates points automatically as the cursor is moved, either one point per unit of time or distance. This mode can easily generate very large data volumes, and should be avoided in most cases. Error correction is especially difficult in this mode. *Point select mode* allows switching between point and stream mode. This mode is sometimes used when lines are both geometric and natural, such as following a straight road and then a river.

At this point the data are ready either for direct integration into the computer mapping software, or are ready to be used as input for cartographic transformations, to change data structure, map base, or scale. As a human-machine interactive process, the digitizing process is only just being studied in detail by cartographers. The process is important to understand, since most of the errors in digitizing can be reduced or eliminated using some simple ergonomic principles (Jenks, 1981).

The second digitizing process is *automated digitizing*. This is also of many types, and indeed is rapidly broadening in scope as a means of data capture. The earliest form of automatic digitizing was the scanner, a device which receives a sheet map, sometimes clamped to a rotating drum, and scans the map with very fine increments of distance measuring the radiance or sometimes transmission of the map when it is illuminated, either with a spot light-source or a laser. The finer the resolution, the higher the cost and the larger the data sets. A major difference with this type of digitizing is that lines, features, text, etc. are scanned at their actual width, and must be preprocessed for the computer to recognize specific cartographic objects. Some plotters can double as scanners, and vice versa.

For scanning, maps should be clean and free of folds and marks. Usually, the scanned maps are not the paper products but the film negatives and scribed materials which were used in the map production. An alternative scanner is the automatic line follower, a scanner which is manually moved to a line and then left to follow the line automatically. The Altek Apache digitizing cursor is halfway between this and semi-automated digitizing, since it is a manual cursor which has a small scanning window on the cursor itself. Automatic line followers are used primarily for continuous lines such as contours. These and other scanners are very useful in CADD systems, where input from engineering drawing and sketches is common.

Increasingly, video scanners are becoming important geocoding devices. These scanners are simply television cameras, sometimes with highly enhanced resolution, which can be mounted on a stand and pointed at a stationary image, air photo, or map. Early versions were monochrome only and had limited gray levels. More recently, color versions with many gray levels, and even color look-up tables, have been used. Map separations can be scanned and entered as separate data layers, and data can be sent directly to a microcomputer with local editing, storage, and even image processing capabilities. Even complex maps such as topographic quadrangles can be scanned by these devices with suitable results, though for a whole quadrangle multiple scans must be used for a reasonable resolution. Again it should be noted that the scanner sees folds, pencil lines, erasures, white-out, and coffee stains as easily as cartographic entities. Great care should be taken with scanning not to introduce complex geometric relationships between the map and the image, such as the effects of different lenses.

Finally, very low cost scanners are now available which can read and interpret both documents and text. This is important, since typed and other text can be entered directly from a map or other document, and then manipulated and plotted within a mapping system. Simple graphics scanning is rarely adequate for cartographic purposes, but can be used to put a rough sketch into a CADD system for reworking. In this way, first-draft or worksheet sketches can be used as the primary source of information for developing the final map design. Most of the graphics in the book were produced using a combination of scanning rough sketches and graphic editing using a CADD-like package.

4.7 TOPOLOGICAL GEOCODING

Bearing the goals of geocoding in mind, let's look at a real geocoding method. More important, let's look at how it has evolved over time. The example we will use is the geocoding associated with the mapping needs of the United States Bureau of the Census, part of the Department of Commerce. As we will see, this geocoding system is particularly good for seeing how topological geocoding became important over time.

The specific mapping need is to support the decennial census effort (as required by the Constitution), by generating street-level address maps for use by the thousands of census enumerators. Fairly early on the use of the computer was recognized as critical. An early system, the address coding guide, was largely a text database, both computerized and manual, that listed all street addresses within census enumeration districts. Included as part of the record was a designation listing the block side of the address, each one given a unique number, as well as geographic information such as census tract, block, ward, post office codes, etc. Eighty-eight of the standard metropolitan statistical areas (SMSAs) for the 19th census (1970) were entered manually, with the rest coming from existing computerized mailing lists and post office checks. All 233 SMSAs were covered by this system, with over 40 million addresses. A need was identified to link this address information with the hand-drawn enumeration maps used in the field. These maps were compiled by asking local communities for all available maps, which were then used as a base for designating the enumeration districts. Understandably, the lack of standardization and variation in accuracy and age of the maps was considerable.

In planning for the 1970 census, the ACG was tested in 1967 for New Haven, CT. Since a major area of interest in this pilot study was computer mapping, an attempt was made to add geocodes to the ACG files. At that time, digitizing tablets were rare, and the Census Bureau actually had to design and build its own prototype in-house system. Many maps were digitized using light tables and graph paper. Geocodes with state plane coordinates were added to the ACG records, although some records used latitude and longitude and "map miles". This is one of the earliest cases of using digitally encoded topology to supplement location geocodes. The process was such a difficult conflict with the errors and inefficiencies of the digitizing methods used, however, that further use of ACG in this context was abandoned.

The resultant technical steering group, which was overseeing the census use study in New Haven, recommended developing new methods for future censuses. A proposal was made to build geographic base files for the census separately, using graph theory as the underlying concept. Each street, river, railroad track, municipal boundary, or other map feature was considered as a straight-line segment. Curved streets were constructed from multiple straight segments. Each node, line segment, and enclosed area was uniquely identifiable within the full network. Line segments were labeled with street names from the base maps, and nodes were numbered sequentially. The entire system was called dual independent map encoding (DIME), because the basic file was created by computing two independent incidence matrices from the source map, line segment/node, and line segment/enclosed area (Department of Commerce, 1970). DIME, therefore, built substantially upon ACG in that it allowed external checking of the logical consistency of the data by performing topological checks. A DIME file was constructed for New Haven. The major difference between this file and the ACG file was that DIME records contained codes for both sides of a street, i.e., the records were segment-rather than block side-based. The first few DIME files were used for experimental computer mapping, using such software as MAP01 and SYMAP.

Between the 1970 and 1980 censuses, DIME was updated and extended in scope. The geographic coverage was also extended, so that a geographic base file of many SMSAs became available to census users. Automated error-detecting methods were devised which used the topological geocodes to locate and correct errors. Names were standardized and header records were devised for the computer tapes on which the GBF-DIME data sets were distributed. For the 1980 census, virtually all the major SMSAs were covered, with a total of 300,000 enumeration districts. The 1980 GBF-DIME files were used commercially, and have remained as important sources of data for a large number of applications.

Analytical flexibility was a key to DIME. Although the geocodes were designed as a way of producing enumeration maps, it was soon realized that DIME was suitable for more general computer mapping. DIME was used for collection of statistical data, automatic generation of centroids, thematic mapping, and automatic address matching, a form of geocoding in itself. Much of this success has been due to the fact that DIME is georeferenced into the census data itself, allowing automatic thematic cartography with thousands of attributes to choose from.

After the use of GBF/DIME for the 1980 census, people started looking at automated thematic mapping as an effective means of understanding the mass of data the census provided. One of the problems for mapping directly from DIME, however, was that most of the segments were street or political boundaries, and that they are highly generalized. The demands of detailed computer cartography from census data called for a revision of the DIME system. As noted above, an important measure of a geocoding strategy is the ability to adapt to demands previously unforeseen. DIME gave a certain storage efficiency and more analytical flexibility than was at first realized. It produced a lot of labor in geocoding, there were errors in the original GBF files, but in the long run the system proved invaluable, succeeding in its original goal and going on to different applications.

For the 1990 census, the Census Bureau developed a system called TIGER, topologically integrated geographic encoding and referencing. TIGER came with a refinement of the DIME terminology (Figure 4.10). Instead of using the block face or street segment as the basic entity, the TIGER system recognizes cartographic objects of different dimensions. The objects are points (nodes), lines (segments), and areas (blocks, census tracts or enumeration districts). In the TIGER terminology, points are zero-cells, one-cells are lines, and areas are two-cells (Figure 4.11).

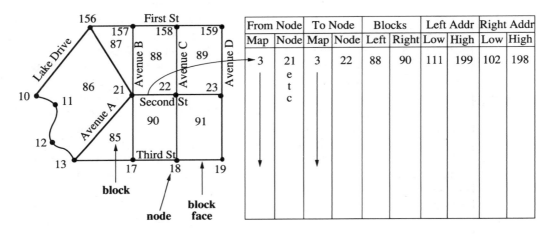

Figure 4.10 Geocoding using the DIME system

Under the TIGER system, the digital cartographic boundaries of census tracts follow real things on the earth's surface. Cartographic elements which were digitized included roads, hydrography, political boundaries, railroads, and some miscellaneous features such as commercial buildings. Each was digitized as a point or a line, and assembled to enclose areas. Interactive checking and maintenance software allowed manipulation of these objects, individually and collectively, as well as labeling and consistency checking. The new files were structured topologically, and linked together by cross-referencing rather than having all data in a single file. Thus linkage with the address records is possible, as is the isolation of the data and map base required for a particular thematic map. Choropleth maps, for example, need only the attributes and the two-cells, while detailed maps need mostly one-cells and labels.

This required a whole new series of maps. A large-scale cooperative effort prepared these maps for the 1990 census. Focal to the effort was having part of the digitizing phase performed in collaboration with the United States Geological Survey. As a result, the Geological Survey was able to convert to a digital basis many of the 1:100,000 series and some of the 1:24,000 series topographic maps.

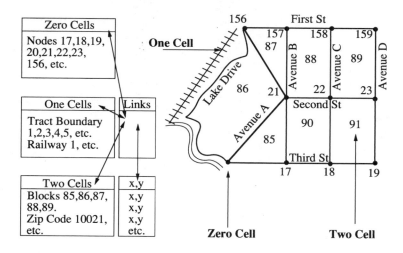

Figure 4.11   Geocoding using the TIGER system

In the evolution of the use of digital data at the Bureau of the Census, foremost is the collection of massive amounts of digital map data. By changing the basic geocoding method, going from a system based on the block face with a sequential topology to a system based on entities of different dimensions, there has been a substantial increase in analytical flexibility. At the same time, intergovernmental cooperation has reduced the labor in geocoding. The incorporation of topology has allowed a substantial amount of automated error checking, which has improved the quality of the data. The evolution from GBF/DIME to TIGER, therefore, has achieved many of the goals for improving geocoding. In addition, the TIGER data are likely to prove an important archive for future map production of many different types and by many different groups, proving once again that success can be measured by the ability to plan for the unanticipated future applications of geocoding.

4.8 REFERENCES

Department of Commerce (1970) *The DIME Geocoding System*, Report Number 4, Census Use Study, United States Bureau of the Census, Washington, DC.

CHRISMAN, N. R. (1987) "Efficient digitizing through the combination of appropriate hardware and software for error detection", *International Journal of Geographical Information Systems*, vol. 1, no. 3, pp. 265-277.

Department of the Army (1967) *Grids and Grid References*, TM 5-241-1, Headquarters, Department of the Army, Washington, DC, AF Regulation 64-4, vol. 1, U.S. Government Printing Office, Washington, DC.

JENKS, G. F. (1981) "Lines, computers, and human frailties", *Annals, Association of American Geographers*, vol. 71, no. 1, pp. 1-10.

LEE, D. AND SALLEE, T. (1970) "A method of measuring shape", *Geographical Review*, vol. 60, pp. 555-563.

PAVLIDIS, T. (1978) "A review of algorithms for shape analysis", *Computer Graphics and Image Processing*, vol. 7, pp. 243-258.

5

Data Storage and Representation

5.1 STORAGE MEDIA

Computer cartography demands a familiarity with the tools and limitations of digital technology. In Chapter 4, we examined the conversion of map data to digital form. Once a map consists solely of numbers, the cartographer is faced with many decisions about how to store the map data, and how to select methods of representation which accomplish the goal of allowing the data to be used for mapping. In Chapter 2 we surveyed the various type of devices in use in computer cartography, especially graphic input and output devices, but explicitly left out storage devices. The ultimate purpose of geocoding and choosing a representational method and data structure for cartographic data is to store the digital map on some kind of permanent storage device. Ideally, the data should be easy to get back from storage, and should also be stored in a way that is both permanent (usable for more than 10 years), and transportable to other computers. Over time, storage technology has drastically reduced the cost and space requirements for storage, and permanence has improved. The key issue for storage remains portability between computers.

For many years, the most widely used storage method was the punched card. A directly comparable medium was paper tape. These storage methods were not permanent, and suffered from additional problems, such as sequencing and tearing. Disk storage replaced cards and paper tape. Fixed or hard disks are usually part of the computer and as such are not transportable, unlike their smaller but removable equivalent, the removable disk. Removable disks, however, are fragile and bulky and can be damaged by strong magnetic fields. Floppy disks, introduced when microcomputers became available, are relatively rigid, so a preferable term in use is *mini-disk*. Older disks were enclosed in a plastic envelope containing the thin magnetic sheet which holds the data.

Early types were 133 mm (5.25 inches) and 203 mm (8.0 inches) in diameter, but are rapidly being replaced by 89-mm (3.5-inch) disks, which have a more solid plastic case and can support higher storage densities.

Probably the longest surviving and most reliable storage medium is magnetic tape, which comes both as loose reels and cartridge tapes. Different lengths of tape, and different numbers of bits per inch (BPI) of storage density, allow different amounts of data to be stored. The cost, reliability, transferability, and high density of magnetic tape make this the normal medium for data and software distribution for large computers.

More recently, mass storage has become possible using optical disk technology. Many optical disks are read-only and work much like a record player, which can play all or part of a record but cannot record. The first generation of optical disks used compact disks as read-only memory, and disk templates had to be written in much the same way that a book is printed. More recent optical media allow both reading and writing. The advantage of optical media is the size of the storage, sometimes approaching gigabytes of storage even for a microcomputer. Almost all topographic map coverage for a single state, both current and historical, would fit onto a single optical disk. The disks themselves, however, are expensive and are comparatively fragile, although prices continue to fall.

Even reliable storage such as disk memory has to be backed up to a more permanent storage medium. Backups are full-system or incremental, in which we just copy every new or changed file on the entire system. All files are copied onto magnetic tape, and then we archive the tapes. If anybody inadvertently deletes a file, or something goes wrong with the computer, we can retrieve things as they were on a given day. Magnetic tape readers are available for microcomputers and are usually used to back up a hard disk on magnetic tape. Even microcomputer hard disks must occasionally be backed up to mini-disks, which are more likely to survive catastrophic events than hard disks.

5.2 INTERNAL REPRESENTATION

The various storage media in use are simply the mechanisms for storing digital cartographic (and other) data. To gain insight into computer cartography, we still have to understand how it is that the data are translated into the physical storage properties of the particular medium, be it a magnetic tape, an optical-disk, or a floppy disk. Using any storage medium involves keeping track of information about where different data are stored on that disk, or tape, or whatever you happen to have. This involves blocking, or dividing the data into manageable chunks, and requires a mechanism for dividing our disk into blocks and sectors. This blocking is independent of any spatial data structure which we may impose upon the digital cartographic data. A floppy disk has radial sectors broken up into blocks or areas reserved for storage of different kinds. The operating system is designed to keep track of this information for you. The link between the actual map of the blocks, the sectors, and the physical disk itself is the directory. The directory is an area of the disk reserved for information about where other things are on the disk. All operating systems have a command to allow you to see what files occupy your storage, although most mask from the user the specifics about where the storage is actually located. Many computer operating systems allow you to have access to much more storage than is apparently available, a method called virtual memory.

Addresses are locations in storage, both permanent and temporary, where information can be stored. The very lowest level of address normally contains the operating system. The basic storage unit is the lowest-level piece of information we can fit into or retrieve from an address. On an optical storage system, addresses relate to resolution. On a mechanical or electrical system addresses are related to the ability to hold a magnetic or electrical charge. Storage devices are state storage mechanisms; they can store on or off, binary 1 or 0, on one bit. A capacitor can hold a charge or not. An optical reader may see a bar code with a black line or a white line. The mathematics that corresponds to this bit logic is called Boolean mathematics, and it works in number base two, the counting mechanism we would use if we had one hand and one finger. Counting is in the sequence ... zero, one. As we string the bits together we can build bigger and bigger numbers. There is a clear map between distribution of digits in binary numbers and what the memory storage actually looks like. On the older storage media, such as cards and paper tape, you could actually see the binary representations as rows of holes punched out of the paper. On magnetic media, the tape either holds a charge or does not.

As an example, the binary number 1010 0011 0001 corresponds to the decimal number 2,609. Notice that just as we break off the 2 from the 609 with a comma for ease of interpretation (most other nations use a space), we use spaces between groups of four binary digits or bits. This has the distinct advantage that it is easy to represent the number in other number bases. The two most common alternative number bases are base eight (octal) and base sixteen (hexadecimal). By far the most used is hexadecimal, in which the counting numbers are 0,1,2,3,4,5,6,7,8,9,A,B,C,D,E,F. Each cluster of four bits can be represented as a single hexadecimal digit. Thus 1010 translates to decimal 10 or hexadecimal A, 0011 translates to decimal 3 or hexadecimal 3, and 0001 translates to decimal 1 and hexadecimal 1. So the binary number above can be represented as the three hexadecimal digits A31. To write the binary equivalent of this value onto the storage medium, many different systems are used. Some write the bit values left to right, some right to left, and some invert the value of the bits byte by byte, a process known as byte swapping. The differences relate to whether the least significant bit (i.e., rightmost logically) is stored first or last.

A cluster of four bits, as can be represented by a single hexadecimal digit, is called a nibble. Two nibbles clustered together are called a byte. We often use the term *word* to refer to the grouping of bytes which the computer itself uses as the basis for storage. Thus we may hear that certain chips are 16-bit, or that a particular computer has a 32-bit word. In the early days of microcomputers, there was a big difference in the size of a word on a microcomputer and the size of a word on a large computer. There were even differences company by company on what the size of a word was on a large computer. Today, even some small computers use the same word sizes as those used by very large computers.

Since hexadecimal is an internal representation of the data and programs that are stored on a computer, many computer users never need know this level of detail. Similarly, we may be able to drive a car without opening the hood. When these details become important is either when something is wrong with the car, or when we decide to build a new car. Programmers make frequent use of binary and hexadecimal, but as a computer user the only time you may see hexadecimal is when something goes wrong. When a program or operating system crashes, it sometimes tries to send a message to assist in finding what went wrong. What the computer sends in the message is an address, or a location in memory where the error occurred, giving the address in hexadecimal.

Some scientific pocket calculators do binary, octal, and hexadecimal arithmetic. Using these calculators is a good way to get some experience in using these number systems, although using them for shopping or balancing a checkbook can be confusing. As well as the usual number operations we can combine binary digits with operators called AND, OR, and NOT. AND simply matches two binary numbers and results in a 1 only if both bits are 1. OR results in a 1 if either bit is 1. NOT simply results in the opposite of a bit, zero replaces 1 and vice versa.

Hexadecimal codes can be used to number the available storage locations in memory, just as an easting and a northing give a location, and a grid row and column number point to a particular pixel. Having found the storage location, however, we may wish to retrieve what we find there, or change the memory to what we want. If the memory locations themselves held hexadecimal digits, then the computer would only be able to store numbers, and only positive numbers at that. To get over this limitation, we have developed a standard way of encoding the more complex sequences that we actually use, numbers, characters such as upper- and lowercase letters, and special characters such as brackets and exclamation points. We don't actually store the letter "e", or the number "0", we store a representation of them in these basic storage codes or building blocks. What gets stored in the memory locations pointed to by hexadecimal addresses are binary strings which correspond to character sets.

A character set normally consists of upper- and lowercase letters, the numerals zero through nine, special characters, periods, dashes and underlines, and characters specific to another language, such as a cedilla, an umlaut, or a circumflex. Fortunately, we have standards for encoding character sets; otherwise, every computer would use its own standard. There are two main standards. IBM decided that it would have its own standard, known as EBCDIC (Extended Binary Coded Decimal Interchange Code). In EBCDIC one character maps into one byte of eight bits. Two nibbles, or eight binary digits, correspond to each character, so that in each location in memory we store a single byte, or eight bits. For example, there are 64 codes within the EBCDIC system. The first four bits are especially important. If they store a hexadecimal C, the stored number following is positive, and if they are hexadecimal D, the number is negative. Either of these two cases mean that the following bits should be taken literally as a number. Hexadecimal 30 through 40 are the special characters in EBCDIC. Special characters C1 through B9 are the uppercase characters, and F1 through F9 are the decimal numbers. Thus in EBCDIC the number 9 is stored as 1111 1001 in binary (F9 in hexadecimal) and negative 9 is 1110 1001 (E9 in hexadecimal).

Generally, the most universal character set is the ASCII (American Standard Code for Information Interchange) set. The ASCII character set stores one character to one byte, giving us 255 possible codes, although only 127 are normally used. The first code is called null, or nothing (not zero). The next codes correspond to "control" a through z. Remember that on a computer the "control" key is just like "shift" on a typewriter, the difference between uppercase a (A) and lowercase a (a) is: hold down the shift and hit a (A), or don't hold down the shift and hit a (a). Control works in exactly the same way as shift. Just as we have upper- and lowercase characters, we have control characters in the code. ASCII control codes go from hexadecimal 01, which is control-a, through hexadecimal 1A, which is control-z. ASCII code 1B corresponds to the "escape" key. Many escape sequences do special things, such as graphics. Many graphics devices when they get ASCII code 1B hexadecimal (27 decimal) do not interpret the next thing they get as a piece of data, they interpret it as a command.

ASCII code hexadecimal 20 (decimal 32) is a space; then we go through all the special characters, exclamation points, pound signs, periods, etc. At ASCII code hexadecimal 30 (decimal 48) we start with numeral zero and go up through ASCII code hexadecimal 39 (decimal 57), which is numeral nine. At hexadecimal 41 (decimal 65) we start with uppercase a, up to hexadecimal 5A (decimal 90), which is uppercase z. At ASCII hexadecimal 61 (decimal 97) we start with lowercase a, and at hexadecimal 7A (decimal 122), we are at lowercase z. Internally, therefore, what actually get stored on the storage medium we are using for computer cartography are the binary representations of ASCII codes, and these values are written into hexadecimally referenced locations in either temporary or permanent memory. The advantages of storing ASCII codes are many, and include being able to view and edit the data, as well as maintaining a high degree of portability between computers. As we will see, however, using ASCII codes often leads to memory management problems.

5.3 STORAGE EFFICIENCY AND DATA COMPRESSION

An important goal for the storage of data within computers is to achieve storage efficiency. We often have to make a trade-off between storage requirements and the ease of use of cartographic data. Ease of use is reflected by the competing goals of achieving rapid access to data, sustaining fidelity, allowing concurrent access to multiple users, and facilitating update and maintenance. In Chapter 4 we noted that one of the fundamental characteristics of geographic and cartographic data is volume, as geographic data sets are typically very large. This means that storage efficiency problems are pertinent to handling cartographic data. Frequently, we are concerned with how we can contort the data to fit into our particular storage device, or to fit our logic or reasoning system. To be storage efficient, we want to retain as much *cartographic information* with as little storage as possible.

Storage efficiency concerns two sets of storage demands. First, the entire digital cartographic data set must reside in permanent storage and be made accessible as required for mapping. Second, a computer cartographic program must bring part or all of the map data into the random access memory (RAM) of the computer for display, analysis, and editing. The computer program moves the data from storage into a data structure supported within the program itself. Data in RAM are available immediately to the program. Data in secondary storage must be brought into and out of the RAM, a process known as I/O or input and output. To move data between these two types of memory is time consuming. I/O is the most time-consuming part of producing maps by computer, since digital cartographic data bases are typically large. When only part of the map can fit into the RAM at any given time, for example on a microcomputer which typically has less than a megabyte of available RAM, memory management can become a significant part of the computer mapping software.

The storage efficiency of the cartographic data already in RAM is determined by the precision of the data and the suitability of the programming data structure. This programming data structure is made up of the physical structure into which the data have been put, and the power of the logical data structure as a construct for solving the particular mapping purpose. The distinction between logical and physical data structures is important, and allows different degrees of storage efficiency to be attained.

In Chapter 4 we considered the merits of various graphic input devices. We noted that there are two major types, semi-automatic and automatic. In general, the semi-automatic devices produce vector data and the automatic produce raster data. Vector data can be very storage efficient, since we only need to capture information when there is information worth recording. For example, on the right map projection we can represent the boundary of the state of Colorado with only four points. Curved lines are sometimes a little less efficient because to maximize the amount of information stored we have to concentrate data in areas where the boundary changes in direction. In digitizing a line we capture more points in areas of extreme curvature than in areas that are very straight. This introduces the concept of information content. What is a high information point and what is a low information content point? Figure 5.01 shows some lines with different degrees of information content at different points. On a straight line, the two end points are very high information points, because they contain information about where the line begins and ends, without which it is impossible to draw the line. A point along the straight line is referred to as redundant, since it adds nothing to the information content. It does, however, add to the data volume. To add a third x and y value along the straight segment, we have to increase the x and y data storage volume by a third, but we get no corresponding increase in the information content. To achieve storage efficiency we seek to minimize the redundancy within the data and to maximize the information content. In a vector system we can vary the sampling density to correspond to what we are mapping.

This concept does not apply only to lines, but to points and areas, too. For example, on a soil map, for which we know that there are two basic geology types with one twice as abundant as the other, we would anticipate two different types of soils. We know that there is a boundary that runs somewhere within the map area. How would we sample that area to achieve the maximum amount of information content in the data versus the minimum amount of redundancy? It would be pointless to put a grid over

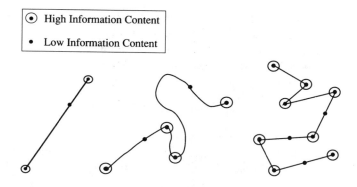

Figure 5.01   Lines with different information content

the map and take soil samples at the intersection of every grid point. Most of our data would be redundant and the sampling strategy would probably miss the only significant feature, the boundary. If we know there is a boundary, we need only a few samples over the map area but many along the boundary itself. This would locate the boundary precisely on the map. Any other sampling strategy would contain redundancy. A vector-based system, in this case irregularly distributed points rather than lines, allows us then to go for the highest information content areas on the map, and concentrate our sampling effort on those. This strategy, however, assumes some prior knowledge about the nature of the geographic distribution involved.

Raster systems work differently. Using a grid, we have one data value for every grid cell, or grid intersection point. This means that we have a large amount of embedded redundancy and high data volumes for the same geographic area. This is the same as uniform sampling. Within a lake, for example, all grid points simply repeat the fact that this is still a lake. This strategy is most suited to geographic phenomena which change continuously, such as elevation and physical properties, rather than discontinuous properties such as surface geology and human characteristics like rural land use. Even so, a grid resolution suitable for one geographic region or for a particular phenomenon with distinct geographic properties is usually unsuitable for another. Thus we would use a different grid spacing for elevations on the Great Plains than in the Rocky Mountains. Grids happen, however, to be convenient for measurement technology. Thus grids are often used for large areas at small scales, while vectors are frequently used for small areas at large scales. The two data structures are really answering two different questions and making two different demands on the data. The two systems are both space sampling strategies, one regular and the other irregular, one with constant resolution and one with variable resolution. If we accept a direct relationship between scale and resolution, then vector and raster conversions are a problem of scale transformation. Tobler has suggested the single term *resel*, or resolution element, to apply to both regular and irregular areal sampling distributions, with the *pixel*, a square or rectangular resel, as a special case.

A distinction between logical and physical data storage is important for storage efficiency. In this chapter we will consider the physical aspects of storage. In the next, we will consider the major logical map data structures. We saw in Chapter 4 that under the UTM coordinate system we could represent the location of almost any point on earth to a precision of 1 meter as a string such as

4,513,410 m N; 587,310 m E; 18, N.

During the digitizing stage of geocoding, at some stage we moved a cursor to this point and pressed a button. What geocode does this process transfer from the map to storage? The full reference consists of 15 digits, plus a binary value for the hemisphere (north or south). The whole string, as shown above, is actually 35 characters, 15 digits, four commas, seven blanks, two semi-colons, five letters, one period, and an end-of-line character to separate this from the next line. Simply storing the fact that the values consist of a UTM reference for zone 18 N as a northing, followed by an easting, reduces the storage to 15 characters:

4513410 587310

By storing a line such as

UTM zone 18 north, northing, then easting.

only once, for 42 bytes (one ASCII code per character), we can save 20 bytes per record. For 100,000 records, we save 2 megabytes at a cost of 42 bytes. Let's see if we can do better than that. Zone numbers go from 1 to 60. When we are designing storage systems, we start by looking at the maximum value we need to fit in any field. We only need to store zone numbers between 1 and 60. For the hemisphere we can store either 1 or 0, one binary digit, instead of a whole byte. What is the maximum easting? If we reached 1,000,000 we would be in the overlap between zones, so for all practical purposes 999,999 is the maximum. The maximum northing is 10 million. Since 10,000,000 south can be represented as 0 north, and since we cut off at 84 degrees north, the maximum northing is 9,999,999. Thus in reality each record can consist of two numbers, one between 0 and 9,999,999 and one between 0 and 999,999. Using hexadecimal, these numbers are 98 96 7F and 0F 42 3F, requiring a total of only six bytes per coordinate pair. Furthermore, if these bytes are written sequentially, i.e., without the new-line characters, our 100,000 records need occupy only 600 kilobytes. The data set, which formerly would have occupied a significant portion of a hard disk, would now fit easily into RAM on a large microcomputer. If we took the additional step of dropping the last digit, i.e., giving our data a precision of 10 meters and storing the first 100-km-grid-square letter and number designation as shown in the Chapter 4, we need store only eight digits, four each for x and y, making only 10 ASCII characters (bytes) per record, or two values with a maximum of hexadecimal 27 0F, making four bytes per record. The 100,000 records now take up 1 megabyte as ASCII codes or 400 kilobytes as binary sequential. By simply economizing on the storage representation of a record, we have reduced a data file from 3.5 megabytes to 400 kilobytes.

This is an example of achieving storage efficiency by physical data compression, reducing the actual number of bytes required to store information. However, the storage improvements achieved here, a compression to 11.4% of the original size, were achieved in two ways. First, we used physical compression, achieving savings mostly by using binary instead of ASCII codes. This compression is directly reversible. Second, we dropped the last digit of the data, saving space but making the compression a non-invertible transformation of the data, actually a scale transformation since we have spatially generalized the data. It is impossible to get this detail back accurately, so a scale change is partly a physical but also partly a logical compression of data storage volume.

The example above is not far from reality. The United States Geological Survey's land use and land cover (LULC) data use a similar compression method. The LULC data in a format known as GIRAS are the digital equivalents of the 1:250,000 land use and land cover maps, available for the whole country. In these data sets, which are also topologically encoded, eastings and the northings are relative to the nearest 100,000 m UTM intersection to the west and south of the map area. This is equivalent to storing once the letter and grid number designations from the military grid, covered in Chapter 4. These are stored separately, once per file. The eastings and northings are then only five digits long. In addition, the minimum resolution is set to 10 meters rather than 1 meter; in other words, they lop off the last digit. In storing an arbitrary origin once, the zone number and the hemisphere are stored also. This leaves eight bytes for each point since the files are ASCII. No new-line characters are necessary since the points file is simply written sequentially.

Often digital cartographic data-bases have differential accuracy levels embedded in them. In these systems, for each x and y value there is another value, an identifier, which says: "include this point at this level". To make a very highly generalized map, we would scan this last character and include the point only if it meets the lowest generalization level, such as 9, which would be the least detailed. A level 9 map would simply exclude all points with level less than 9. A level 1 map would include every single point. Using this method, we can embed different levels of resolution within the same data set, at the cost of storing one more character per point. An effective way to determine the appropriate resolution factor is to apply a line generalization method, such as the Douglas-Peucker technique, discussed in Chapter 10.

Many data compression methods are available to all data files, whether spatial or not. A popular method of physical data compression is Huffman coding. Huffman coding puts an entire file through a one-step process, the result of which can be a significant saving in storage. The process is entirely reversible, and in doing so the reconstructed file is exactly the same as the original. Files with much redundancy in them, many blank lines and columns for example, or large numbers of similar digits, compact significantly with Huffman coding. This is often the case with map coordinate files. Huffman coding works on ASCII files and performs a frequency analysis of the ASCII codes. The ASCII codes are then replaced by a a new set of Huffman codes of variable length, with the very shortest-length codes corresponding to the most frequently-occurring ASCII codes (Held, 1983).

Necessarily, there is a substitution. To get storage efficiency we have to sacrifice something, and what we usually sacrifice is ease of retrieval, or analytical flexibility. Especially now that the cost and limitations of data storage are falling, the savings of space may not be worth the loss of flexibility or precision used to achieve storage efficiency. As a rule, ASCII data representation is superior for cartographic data unless the mapping is in a production environment, since errors can be detected simply by looking at the data with an editor. The human eye and brain is a powerful error detector even for large numbers of records, especially when coupled with automatic error-detection routines. While data compression may be desirable for data transfer through networks, or archiving on tapes, cartographic data are best represented as the actual location coordinates they ultimately must consist of.

Given the basic means by which data can be stored and compressed on computer storage media, which methods are used in computer cartography, and why? In the following section, we will examine some of the formats which have been developed and used, especially by government agencies, and determine the significance of the effort to advance a national digital cartographic data standard.

5.4 STANDARD DATA STORAGE FORMATS FOR CARTOGRAPHY

The beauty, and also the problem, of standards is that there are so many to choose from! Since most digital cartographic data archives were developed piecemeal over a long period of time and for an enormous number of applications, there are a great many proposed standard formats. The need for standardization is most apparent when data must be transported between applications and between computer systems. The most successful data formats are those which have withstood transportation between systems, and those which have had the most use and documentation. In this section, we will examine in detail the major producers of digital cartographic data from the federal government—the Defense Mapping Agency (DMA), the National Ocean Service (NOS), the United States Geological Survey (USGS)—plus a widely distributed data format, the World Data Bank. The logic of the Bureau of the Census's formats was discussed in Chapter 4. Following this, we will examine the digital cartographic data standards and how they relate to the future provision of digital map data.

5.4.1 Standard Formats at the Defense Mapping Agency

The Defense Mapping Agency (DMA) is the primary producer of digital cartographic data products for support of the United States military. The digital data form part of the Digital Landmass System (DLMS) and are made up of two parts, the Digital Feature Analysis Data (DFAD) and the Digital Terrain Elevation Data (DTED). Together, these data sets consist of about 25,000 magnetic tapes at 1,600 BPI, structured sequentially. By 1995, this data set will consist of 10^{19} bits of data. Overall, the DLMS contains information about terrain, landscape, and culture, to support aircraft radar simulation, map and chart production, and navigation.

The DTED data contain elevation data sets in latitude, longitude, and elevation format which are used to generate elevations at 3-arc-second intervals. These data are referenced to mean sea level and are integer elevations rounded to the nearest meter. These data have been produced from contour maps and from stereo air photos, and are enhanced with stream bed and ridge information to preserve the topographic integrity.

The DFAD data contain descriptions of the land surface in terms of cultural and other features, forests, lakes, etc., stored as point, line, and area data. These data are lists of or single latitude, longitude pairs, and have accompanying tables of attribute information. For example, a water tower would be stored in DFAD as a point, and the attributes would include feature height and structure type, while a bridge may be recorded as a line segment. Linkage between the attributes and the cartographic data is maintained by header files.

The land features are coded at two levels. Level I data, with a resolution of about 152 meters (500 feet), contain large area cultural information such as surface material category, feature type, predominant height, structure density, and percentage roof and tree cover. Cultural features are digitized in a planimetric linear format, often called standard linear format (SLF). Standard linear format has been suggested as an internal digital cartographic data exchange format for DMA.

Level II data are far more detailed, showing small area cultural information, and match DTED data at one arc second resolution, about 30 meters (100 feet). These data are produced by manually digitizing large numbers of maps, charts and air photos. The level I data-bases cover about 24 million square nautical miles, and will be completed worldwide some time between 1990 and 1995. Level II data are digitized only for a limited number of areas of interest.

5.4.2 Standard Formats at the United States Geological Survey

Digital cartographic data from the United States Geological Survey is distributed by the National Cartographic Information Center as part of the National Mapping Program. USGS digital data falls into four categories. These are digital line graphs (DLGs), digital elevation models (DEMs), land use and land cover digital data (GIRAS), and digital cartographic text (Geographic Names Information System, GNIS). The USGS hopes to complete coverage of the United States by 1993, and will distribute the map products on computer tape, floppy disk, and potentially, optical disk media.

The DLG data are digital equivalents of the 7.5- and 15-minute USGS sheet maps, as well as the more generalized 1:100,000 maps and the National Atlas state-level 1:2,000,000 maps. The DEM data are land surface elevations, at both 1:24,000 with a ground spacing of 30′ meters under UTM, and 1:250,000 with a ground spacing of 3 arc seconds, the same as the DTED data described above. The land use and land cover data are digital versions of the 1:100,000, and 1:250,000 land use and land cover and associated maps (the interim land cover mapping program for Alaska). Finally, the GNIS is a unique data set, which includes the text with its attributes from a large number of USGS map products.

As their name suggests, the digital line graphs are vector data sets and are separated by scale, the scales being 1:24,000, 1:100,000, and 1:2,000,000. At the 1:24,000 scale, the data consists of boundaries (political and administrative, such as the national parks), hydrography (standing water, flowing water, and wetlands), the boundaries of the public land survey system (the township and range system), transportation (including roads, trails, railroads, pipelines, and transmission lines), and other significant man-made structures, such as built-up areas and shopping centers. These maps were digitized from 7.5- or 15-minute quadrangles, or their compilation products. The 1:100,000 data sets are similarly structured, and are subdivided by groups of files covering 30- by 30-minute blocks, from the 1:100,000 scale topographic maps. The 1:2,000,000 maps were digitized from the 1970 National Atlas of the United States of America, and follow the same DLG format. A total of 21 digital map sheets cover the United States, 15 for the coterminous states, 5 for Alaska, and 1 for Hawaii. Coverage for New York, for example, includes all the northeastern states, from New York to Maine.

While these data were originally to consist of three levels of accuracy and coding, in fact all data are provided to the maximum level, DLG-3. DLG-3 data are topologically structured and consist of nodes, lines, and areas structured in a manner that explicitly expresses logical geometric relationships. The coded geographic properties are adjacency and connectivity. This allows both plotting of the data for computer cartography and the use of derived information in analytical cartography. The coordinate system is local to the map in thousandths of an inch, but parameters for conversion to UTM are provided in the header files.

The basic cartographic objects in the files are nodes and lines, with areas consisting of labels and links to the other objects. Lines begin and end at a node, and are geocoded with the left- and right-hand areas they divide. Lines connect at nodes, and no line crosses itself or another line. Islands are lines which connect at a dummy node, and are termed degenerate lines. Areas are delimited by lines and contain a point, to which is assigned a label for the area. In addition, lines have attribute codes which determine what cartographic entity is being represented. These codes are based on the standard USGS 7.5-minute map series symbol list. Figure 5.02 shows a typical segment from a DLG and gives examples of different attributes.

The digital elevation Models are digital maps of the surface elevation of topography. Two scales are involved, 1:24,000 and 1:250,000. The 1:250,000 data are 3-arc-second increments of latitude and longitude for 1-degree blocks covering most of the United States. These data are very similar to the DMA's DTAD data and in fact are derived from them. Each 1-degree block contains 1,201 elevation profiles, with 1,201 samples in each profile. Elevations are in meters relative to mean sea level, and latitude/longitude uses the 1972 World Geodetic System datum. Figure 5.03 shows how this data set is logically organized. These sets are distributed on computer tape, with standard headers and structures.

The 1:24,000 data correspond to the 7.5-minute quadrangles and are in UTM coordinates. As such, the boundaries of the map are not square, and the number of elevation samples varies by south-to-north profile over the map. Each profile has a local datum, given in a profile header at the start of each string of samples. The files are

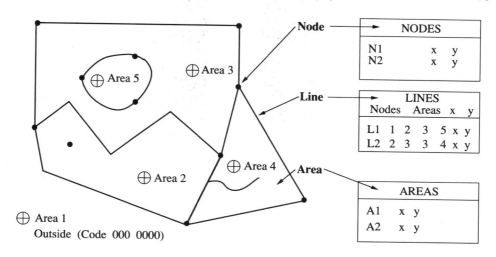

Figure 5.02   Sample DLG data coding format

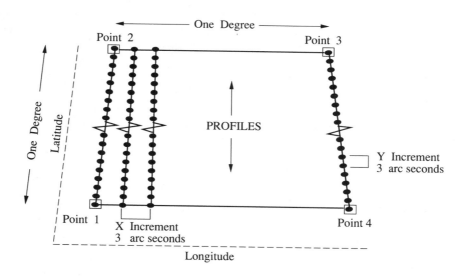

Figure 5.03   1:250,000 3-arc-second DEM format from the USGS

ASCII records, stored sequentially on magnetic tape. Figure 5.04 shows this logical format, and it is important to note that the physical structure depends somewhat upon which side of the central meridian for the UTM zone the data set falls.

The last USGS data set to be considered here is the digital land use and land cover data. These data sets are available by 1-degree by 1-degree block for the entire country, and were digitized from the land use and land cover sheet maps, dating back to

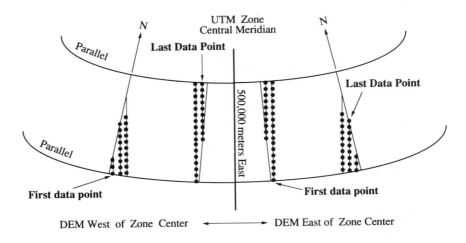

Figure 5.04   1:24,000 30-meter DEM format from the USGS

the late 1970s. Each 1-degree block is split into sections to make files of manageable size; for example, the Albany, NY, 1-degree quadrangle has 15 sections. Data within blocks are stored in GIRAS (geographic information retrieval and analysis system) format. Since the maps consist simply of land cover polygons with their attributes, most of the data are nodes, lines, and links between them to create polygons, a loose topological structure. The map and section headers are followed by an arc records file, which consists of pointers (sequential links) into the next file, the actual coordinates. The coordinates are in UTM and are truncated by rounding to the nearest 10 meters and by using the nearest 100,000 grid intersection.

5.4.3 The World Data Banks

The CIA World Data Banks are widely distributed and used for many mapping purposes. They consist of decimal latitude and longitude pairs. World Data Bank I at 1:12,000,000 consisted of 100,000 x,y coordinates, and when stored sequentially in binary fits onto about 1 meter of magnetic tape. World Data Bank II was digitized at 1:3,000,000, and has about 6 million coordinates. Neither data-base has a topological structure, however. World Data Bank II has several topological inconsistencies. Each major coastline simply continues as a single line until it meets a significant point, such as a major national boundary, or closes on itself. The same is true of the political and hydrological data. The beginning of a new line is the only break from the sequence of binary-encoded pairs of latitude and longitude in degree-minute-second format.

5.4.4 Standard Formats at the National Ocean Service

Under the National Ocean Service falls the Office of Charting and Geodetic Services, which is also part of the National Oceanic and Atmospheric Administration (NOAA).

This office produces two major types of cartographic products, nautical charts depicting hydrography and bathymetry, and aeronautical charts which consist of visual flight charts and instrument flight charts. Although this division of the federal government is comparatively new to the large-scale production of digital cartographic data, some significant data sets are available.

The Nautical Charting Division of NOS offers digital shoreline data and some aids to navigation. These data sets cover the shorelines of the coterminous United States, Puerto Rico, the U.S. Virgin Islands, and Hawaii, all to nautical chart accuracy standards. The data are available at small or large scale. The large-scale data have been digitized from the largest-scale NOS nautical charts of each part of the coastline, and show by vectors the mean high water mark. The shoreline vectors are continuous lines, but are broken where rivers flow into the ocean and where the shoreline cannot be precisely located. Each data set consists of between 100,000 and 300,000 points, and has been edge matched between data sets. Nineteen data sets cover the areas mentioned above, with approximately another 20 planned for Alaska. Three small-scale sets—East Coast, Gulf Coast, and West Coast — will be supplemented with an additional seven for the Great Lakes and Alaska.

The small-scale data are digitized from charts at 1:250,000, and show more generalized shorelines as a result. Records are 80 ASCII characters per record, written sequentially, with each point using one record. Each point is recorded as degrees, minutes, and seconds (to one thousandths) of latitude and longitude, with full geodetic datum, scale, source, date, and feature code reference. In addition, separate files contain point data aids to navigation, such as the locations of wrecks, fixed and floating navigation aids, and obstructions. These data are stored in a similar format to the shoreline data, and contain about 30,000 points per data set. Between the two data sets, just about all the data for the nautical charts are geocoded and distributed.

One final and notable digital data set supplied by NOS is the global elevation data, available from the National Geophysical Data Center, part of NOAA. These data are surface elevations and ocean depths, at 10 minutes of increment for latitude and longitude, digitized from navigational and aeronautical charts by the U.S. Navy. The entire data set contains 2.3 million observations, and is distributed both as the whole world and as segments by continent. The distribution medium is nine-track magnetic tape or floppy disk.

5.4.5 Existing Standards

While many different organizations, both private and public, have developed their own standard formats for digital cartographic data, the disparities between standards have become disadvantageous only since the ability to share and distribute these data has become possible. Designing a system to accept data from two standards, for example, needs either a standards translator or two translators into a third standard. A system to accept data from 10 standards, however, must have either 10 times nine translators or 10 forward and 10 reverse translators to and from a single, universal standard. In the long run, clearly a commonly accepted standard is most desirable. There have been some pioneering efforts to standardize digital cartographic data formats. Among the standards are several developed for data exchange within agencies. These include the Committee on Exchange of Digital Data (International Hydrographic Organization), the Federal Geographic Exchange Format of the Federal Interagency Coordinating Committee on Digital Cartography, the Geographic Data Interchange Language of the Jet Propulsion Laboratory (NASA), the Standard Digital Data Exchange Format of the National Ocean Service, plus those discussed above (Langran and Buttenfield, 1987).

Out of these standardization efforts grew the National Committee on Digital Cartographic Data Standards, more commonly called the Moellering Committee, which was formed in 1982 under the auspices of the American Congress on Surveying and Mapping as a result of a request from the USGS. In 1987 this committee completed the draft stage for a new set of common data standards for digital cartographic data. These standards, which eventually will be submitted to the National Bureau of Standards for submittal as a federal information processing standard, are likely to — indeed already have — substantially influenced computer cartography. As such, they will form the remainder of the discussion for this chapter, concluding with a set of C language programming data structures which are consistent with the standard.

5.5 THE DIGITAL CARTOGRAPHIC DATA STANDARDS

A significant step in the standardization of the treatment of digital cartographic data came with the publishing in 1988 of "The Proposed Standard for Digital Cartographic Data" as a special issue of *The American Cartographer*. In this volume, the history of the standard is reported, and the roles of the organizations and individuals who assisted with the standard are reported. The proposed standard consists of four sections, the definitions and references, spatial data transfer, digital cartographic data quality, and cartographic features. The definitions and references are a systematic attempt to define a set of cartographic primitives as zero-, one- and two-dimensional objects, with which digital cartographic feature representations, i.e., symbols and maps, can be built. It should be noted that the standard definitions are restricted to these object definitions, and do not propose any particular method or type of symbolization. The language of a feature, an entity, and an object is used — concepts which are used throughout this book.

The spatial data transfer component of the standards attempts to meet the requirements for moving digital cartographic data between systems. The USGS will serve as the maintenance authority for this standard, which proposes a series of steps to be taken to attain conformance. The standard makes a large number of specific requirements for data transfer, and provides model formats for the structuring of all types of cartographic data. Transfer forms include global, data quality modules, vector form modules, relational form, and raster form. Clearly, the ability of specific computer cartographic systems to accept raw input data in one or more of these formats is essential if the large amounts of digital cartographic data from the sources mentioned above are to exploited.

The section on data quality proposes that digital cartographic data include a quality report, either as a paper document or as part of the data set. Elements of the report include lineage of the data, positional accuracy, attribute accuracy, logical consistency, and completeness. These requirements seem critical to the long-term survival of data sets, especially within GISs, and also are of critical importance for the provision of an assessment of reliability for a particular map product made from digital cartographic data.

The fourth and final section of the standard relates to cartographic features. Central to the standard here is a complete set of cartographic entity and attribute definitions. This set of features in intended to be as complete as possible, though provisions for update are made. This means, for example, that when a digital data set references a "water tower", there is complete agreement on what constitutes a water tower. Some 213 pages of entity and 34 pages of attribute definitions are included in the standard. For example, the cartographic entity *church*, defined as "a building for public worship, especially Christian worship", is included term number 198 for the entity *building*, defined as "a permanent walled and roofed construction", with standard entity term number 20.

Two other important sections of the proposed standard document are of direct interest. Under the cartographic objects section, a glossary is included. Also, the microfiche inserts at the end of the standards include a remarkably thorough bibliography, to which the reader is referred for additional information.

5.6 C LANGUAGE CARTOGRAPHIC DATA STORAGE

The Digital Cartographic Data Standards have provided cartographers with a standard set of terminology and concepts around which data structures can be developed. While most data structures are expressed in published form as logical models of data organization, in computer cartographic software they find their actual implementation as computer programming data storage mechanisms, what we will call here program data structures. These program data structures must relate to the goals of geocoding discussed in Chapter 4, must relate physically to the actual methods of storage and representation of digital data discussed in this chapter, and also ideally should be simple and effective to use in a programming environment.

Many of the non-structured computer programming languages did not support sophisticated data structures. More recent languages do support these structures, especially the languages most likely to be used in a large number of computing environments, Pascal and C. In Pascal, the *record* is similar to the *structure* in C, and the advanced student will be able to translate most of the computer programs discussed here between the languages with relative ease.

In the C programming language, the *structure* is defined in the following way.

Function 5.01

```
/* Example of a structure declaration in C     kcc   12-88 */
struct example {
float first; int second; };
```

To use this structure in a program, first a variable must be declared to have this particular *structure*.

Function 5.02

```
/* Example declaration of a variable to be of type structure
      kcc            12-88 */
struct example a_structure;
```

Each structure element is then used by linking sub-elements with periods. For example, to assign a value to the first element and print the value of the second, assuming they exist, the following program lines could be used.

Function 5.03

```
/* Example use of data stored in a structure    kcc   12-88 */
a_structure.first = 3.1415927;
printf("The second element of the structure is %d\n",
      a_structure.second);
```

A structure can be multi-dimensional, as in the following example. FORTRAN programmers should note that C language arrays all start at a count of zero.

Function 5.04

```
/* Example declaration of a multidimensional variable to
      be of type structure kcc      12-88 */
struct example {
float first; int second; };
struct example a_structure[5];
for (i = 0; i < 5; i++) {
     printf("Structure element one sub %d = %f",i,
          a_structure[i].first);
     printf("Structure element two sub %d = %d",i,
          a_structure[i].second);
}
```

The final concept for C language structures is that a structure can be part of the declaration of a structure. An example of this follows.

The use of structures permits a high degree of organization in computer programs, especially since structures can be passed as arguments between functions, or declared outside the function definitions to make them available anywhere within a program. To achieve this characteristic, somewhat similar to using COMMON blocks in FORTRAN, the structure definition can be defined en masse in a separate file. C supports such declarations, using the term *header file* for these files. In the following declarations, it is assumed that there exists a file called cart_obj.h which consists of declarations of the C language data structures to be used in this book. In each case, the structures are given upper-case names, so that when variables are declared to have this type of structure, they can use the same name in lowercase without confusion. For example, the first declaration defines the zero-dimensional primitive cartographic object POINT. To declare a variable to be of type POINT, we can still call it *point* (lowercase), and perhaps give it multiple instances. Thus a collection of points, a maximum of 50 for example, could be stored and used in a C language structure *point* in the following way.

Function 5.05

```
/*  Example use of the POINT data structure  kcc   12-88   */
/* The first statement states that the structure definitions
      are in cart_obj.h in this directory */
#include "cart_obj.h"
/* Now declare a structure point, with 50 elements of type
      POINT */
struct POINT point[50];
/* Print, for example, the 30th elements of point
      (indexed by 29) */
printf("(x,y) of 30th element is (%f,%f)\n",point[29].x,
      point[29].y);
```

Since the digital cartographic data standards divide the major cartographic object definitions into zero, one, and two-dimensions, the same subdivision will be followed for the C language structures.

5.6.1 Zero-Dimensional Cartographic Objects

The primitive object with one-dimension is the point (Figure 5.05). Since the x and y values which define a point must store both integer and other coordinates such as UTM eastings and northings, and also global coordinates in both latitude/longitude and radian format, the C basic type of float is used. In some applications it may be preferable to alter these to integers, although careful attention should be paid to whether negative

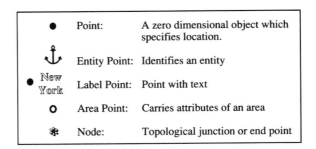

Figure 5.05   Zero-dimensional cartographic objects

values are appropriate, what the maximum values are likely to be, and the loss of precision involved.

Three special cases of a point are given, allowing the point to store an attribute associated with it, either a feature, a label, or an area identifier. The final zero-dimensional object is the node, which contains both a point and linkages to one-dimensional objects connected to the node. In the following part of the cart_obj.h header file, the initial set of statements (the #defines) allow the preprocessor to assign any constants to be used in the structure declarations. For example, instead of allowing a fixed value of 20 as the maximum number of points, the constant MAXPTS is used, and is simply switched at compile time for the constant 20 by the C preprocessor.

Function 5.06

```
/* -------------------------------------------------------
/*   C language cartographic data structures after the
/*   digital cartographic data standards      kcc 5-88
/* -------------------------------------------------------
*/
/* Define the constants */
#define MAXPTS    20
#define MAXROW    20
#define MAXCOL    20
#define MAXHOLE       3
#define MAXSTRING 100
#define MAXARC    100
#define MAXLINK   100
#define MAXCHAIN  100
#define MAXLABEL  20
/* Zero dimensional objects */
struct POINT { float x; float y; };
struct ENTITY_POINT { int feature_type; struct POINT; };
struct LABEL_POINT { char label[MAXLABEL];
     struct POINT point; };
struct AREA_POINT { int polygon_id; struct POINT point; };
struct NODE { int node_identifier; struct POINT node; };
```

An example of a point has already been given above. An entity point may contain a feature type of "church", identified in the feature code section of the data standards as entity type 20 included term 198, for example at 1 degree 30 minutes east, 55 degrees 20 minutes and 45 seconds north. The following declaration and assignment would store this information.

Function 5.07

```
/* Program segment to demonstrate an entity point structure */
struct ENTITY_POINT a_church;
a_church.feature_type = 20198; /* Feature type for a church */
a_church.point.x =  1.3000;   /* Note DDD.MMSS Global coord.
    format */
a_church.point.y = 55.2045;   /* Note DD.MMSS Global coord.
    format */
```

Similarly, the same church may have a specific name to be placed next to a symbol for a church on a map, such as "St. Paul's Church". In this case, a label point should be used.

Function 5.08

```
/* Program segment to demonstrate a label point structure */
    struct LABEL_POINT saint_pauls;
    /* Since the following is an address, the string is
        stored starting
    /* at saint_pauls.label[0], and terminated with a NULL */
    saint_pauls.label = "St. Paul's Church";
    saint_pauls.point.x =  1.3000;   /* Note DDD.MMSS Global
        coord. format */
    saint_pauls.point.y = 55.2045;   /* Note DD.MMSS Global
        coord. format */
```

In the case of the area point, the identifier is a pointer to a polygon which is associated with the point. For example, the pointer could be to a polygon defining the boundaries of the church parish. Similarly, the node simply defines the location and gives an identifier so that the node can be found by other, higher-dimensional objects.

The first encounter within a computer program in the C language with these structures is when the data are read from files. We will use ASCII files throughout this discussion, although very much faster input and output to files is possible if binary reads and writes are used. As noted above, ASCII data files can be edited, examined, modified, and corrected with the use of your favorite text editor. A function to read a single file containing a large number of point features with their labels follows. Assume that the file is called point_data and consists of the following.

Function 5.08

```
File:    point_data
Format: ASCII

  1.3000 55.2045 St. Paul's Church
  2.1023 56.1012 Water Tower
 -1.1213 57.1123 Bench Mark (45.4 meters)
 -1.1557 55.5927 Radio Tower
  2.1519 54.1314 Radar Reflector
```

Then the following function will load the data into the structure entity_point.

Function 5.09

```
/* Read a file of latitudes and longitudes in DDD.MMSS
   format with point labels and load the data
/* into a structure of label points        kcc        12-88 */
#include <stdio.h>
#include "cart_obj.h"
int read_label_points()
struct LABEL_POINT label_point[MAXPTS];
{ int i=0, number_of_points; FILE *file_pointer;
/* Open the input file and abort if it doesn't exist */
if ((file_pointer = fopen("point_data","r")) == NULL) {
    printf("Unable to open file 'point_data'\n"); exit(); }
/* Read the point data, counting the entries as we read them */
while (fscanf(file_pointer,"%f%f%s",&label_point[i].point.x,
    &label_point[i].point.y, label_point[i].label) != EOF) i++;
number_of_points = i;
/* Close the file */
fclose (file_pointer);
/* Return the number of points */
return (number_of_points);
}
```

The data could then be read within a program with the following code segment.

Function 5.10

```
/* Program to use read_label_points() */
some_function()
struct LABEL_POINT label_point[MAXPTS];
{
<do something here>
/* Read the data */
number_of_points = read_label_points();
for (i=0;i<number_of_points;i++) {
    <Do something with each point>
}
<etc>
}
```

5.6.2 One-Dimensional Cartographic Objects

The one-dimensional cartographic objects are more various, serving a large number of cartographic needs (Figure 5.06). The generic name for a one-dimensional object is a line, which can consist either of a locus of points defined by a function such as a polynomial or a b-spline (an arc), or as a sequence of connected primitive objects known as line segments. The line segment is simply a straight line connecting two points, and we define it to be simply two points stored together. The string, a group of connected segments, without any other topological information, makes up the building block for the more complex one-dimensional objects. When a string closes upon itself, it is termed a ring, and rings can be created from strings, arcs, links, or chains. The link is a network primitive, consisting of a single edge of a connected graph. The directed link adds the direction of connection, and the addition of a string between the nodes forms a chain.

This is a complex group of objects, and there is a need to break the list down. A simple division is by those objects which store topology, and those which are designed largely simply to reflect geometry. The geometric objects are the line segment, the string, and rings of arcs and strings. Geometry only is adequate for many plotting purposes, but topology is required for some data structure conversions and analytical operations.

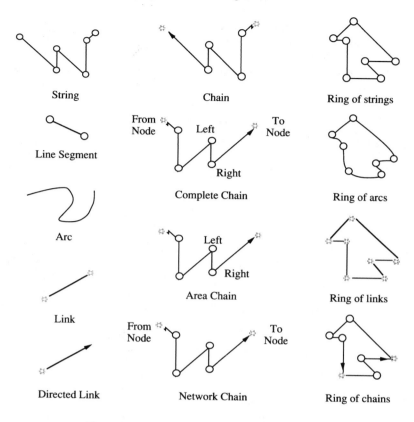

Figure 5.06   One-dimensional cartographic objects

Function 5.11

```
/* One-dimensional objects - Geometry Only */
struct LINE_SEGMENT { struct POINT point[2]; };
struct STRING { int number_of_points;
    struct POINT point[MAXPTS]; };
struct ARC { int arc_identifier; struct STRING string;
    int *function; };
struct RING { int ring_identifier; int number_of_points;
    struct POINT point[MAXPTS]; };
struct RING_OF_STRINGS { int number_of_strings;
    struct STRING string[MAXSTRING]; };
struct RING_OF_ARCS { int number_of_arcs;
    struct ARC arc[MAXARC]; };
```

We define a special structure called a ring, consisting of a single string which closes on itself, since this data structure (the polygon list) is more frequently encountered than the ring of strings. A ring of arcs is rarely encountered, with the possible exception of routines for automated contour drawing. The ring of strings is included for consistency. The line segment as a cartographic primitive is not frequently used. The primary use of this object is in computation, for example of boundary intersections during polygon overlay. The two points forming the segment, in this case, would invariably come from two consecutive points in a ring or in a string. The following code segment illustrates this.

Function 5.12

```
/* Select successive line segments from a string */
struct RING ring;
struct LINE_SEGMENT line_segment;
for (i=1;i < ring.number_of_points;i++) {
     line_segment.point[0].x = ring.point[i-1].x;
     line_segment.point[0].y = ring.point[i-1].y;
     line_segment.point[1].x = ring.point[i].x;
     line_segment.point[1].y = ring.point[i].y;
     <Do something with this segment>
}
```

To read a string, assume a file containing data in the format of one string per line, with the first record being the number of points in the string (n), and the next records being the coordinate pairs, as x1,y1,x2,y2,x3,y3 up to xn,yn. Similarly, a ring is identical to the string except that the first and last point x and y values match exactly, and that the ring cannot cross itself.

Function 5.13

```
File: sample_string
Type: ASCII

 7   0.0 0.0 1.0 2.0 2.0 1.5 2.5 3.5 3.0 3.0 5.0 3.1 7.0 1.0
```

The following main program and function reads this string into a structure of type string.

Function 5.14

```
/* ============================================================
/* Read a string from file sample_string into structure string
/* ============================================================ */
#include <stdio.h>
#include "cart_obj.h"
main() {
FILE *file_pointer;
struct STRING string;
int read_a_string(), i;
/* Open the file, abort if nothing is here */
if ((file_pointer=fopen("sample_string","r")) == NULL) {
    printf("Unable to open file 'sample_string'\n"); exit(); }
if (read_a_string(file_pointer)) printf("String %d has %d
    points\n",++i,string.number_of_points);
/* Close the input file */
fclose(file_pointer);
return;
}
```

The function read_a_string() reads a string into a structure of type STRING and assumes that the header file cart_obj.h both defines STRING and declares string to be of type STRING.

Function 5.15

```
/* ==========================================================
/* Read from an open file a string as a STRING structure
/* ========================================================== */
#include <stdio.h>
#include "cart_obj.h"
int read_a_string(infile)
FILE            *infile;
{
int             j;
/* Read the number of points first, if none, this is the
    end of file */
```

```
if (fscanf(infile, "%d", &string.number_of_points) != EOF) {
    for (j = 0; j < string.number_of_points; j++)
        fscanf(infile, "%f%f", &string.point[j].x,
            &string.point[j].y);
    /* String complete, read the end of line character */
    fscanf(infile, "%*1c");
    return (1);
} else
    return (0);
}
```

The same function will work for a ring, but a check should be implemented to ensure closure. Also, the ring identifier should be read first. To read a ring of strings, the function should be modified to read first the number of strings, then to repeat the input from the function as many times as necessary to read all the individual strings. Again, the first and last points should be matched to ensure closure of the ring.

The remainder of the one-dimensional objects are designed to store topology and geometry. These are the link, the directed link, the ring of links, the ring of chains, the complete chain (called simply a chain here), the area chain, and the network chain.

Function 5.16

```
/* One-dimensional cartographic objects - Geometry and
    topology */
struct LINK { struct LINE_SEGMENT link; };
struct DIRECTED_LINK { char link_identifier; struct NODE from;
    struct NODE to; };
struct CHAIN { int chain_identifier; struct DIRECTED_LINK node;
    struct STRING string;
        int left_polygon; int right_polygon; };
struct RING_OF_CHAINS { int number_of_chains;
    struct CHAIN chain[MAXCHAIN]; };
struct RING_OF_LINKS { int number_of_links;
    struct LINK link[MAXLINK]; };
struct AREA_CHAIN { struct STRING string; int left_polygon;
    int right_polygon; };
struct NETWORK_CHAIN { struct DIRECTED_LINK node;
    struct STRING string; };
```

The simplest of these is the link. A link consists simply of two nodes, stored here as a line segment since order is not important. The four points making up the link are shown in Function 5.17.

Function 5.17

```
link.point[0].x
link.point[0].y
link.point[1].x
link.point[1].y
```

The directed link is a little more sophisticated, and has an identifier, plus two nodes in sequence. A very common example of the directed link is to show traffic flow. A traffic flow network (Figure 5.07) has a series of links, each of which has an associated direction as well as *from* and *to* nodes. The example given is traffic flow around Hunter College, which is surrounded by one-way streets. A data file containing directed links may look as shown in Function 5.18.

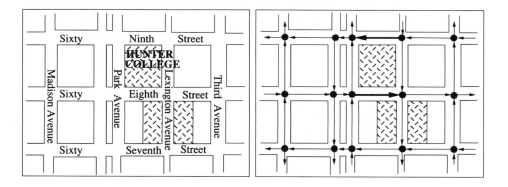

Figure 5.07   A typical directed links map

A program segment to read this type of data may look as shown in Function 5.19.

5.6.3 Two-Dimensional Cartographic Objects

Two-dimensional objects under the standards are generic, with the exception of the pixel and grid cell, which are for graphics only. The *area* is a bounded two-dimensional object which may or may not include the boundary. If the boundary is excluded, the term used is *interior area* (Figure 5.08). The *polygon* is defined as an area consisting of an interior area, one outer ring and zero or more nonnested inner

---

Function 5.18

```
File: sample_directed_links
Format: ASCII

68thSt_Between_Park_and_Lexington  123  587123.0 4513400.0
    587162.0 4513392.0
69thSt_Between_Park_and_Lexington  124  587165.0 4513445.0
    587125.0 4513449.0
```

---

rings. Two sub-types of polygon are the simple polygon, which has no inner rings, and the complex polygon, which has one or more inner rings.

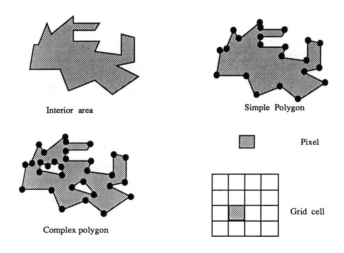

Interior area

Simple Polygon

Pixel

Complex polygon

Grid cell

Figure 5.08   Two-Dimensional Cartographic Objects

The two final two-dimensional cartographic objects accord well with the GKS standard for graphics, and are designed to facilitate the handling of raster and grid data. The *pixel* is a picture element consisting of the smallest nondivisible element of an image. The grid cell is the final object, and is an element of a regular or nearly regular tessellation of a surface. Thus for DEM data, for example, the grid cell is an elevation observation spaced 30 meters apart from its neighbors on a square tessellation, while multiple pixels could be in use in a digital map to symbolize the grid cell. The two-dimensional objects can be represented by the following C language programming data structures.

Function 5.19

```
/* Read a directed link from file sample_directed_links into
      structure d_link
#include <stdio.h>
#include "cart_obj.h"
int read_directed_links ()
struct DIRECTED_LINK d_link[];
{
FILE *file_pointer;
char c;
int i = 0, number_of_directed_links;
/* Open the file, abort if nothing is here */
if ((file_pointer=fopen("sample_directed_links","r")) == NULL) {
      printf("Unable to open file 'sample_directed_links'\n");
      exit();
      }
/* Read the whole file */
while( c = fscanf(file_pointer, "%s", d_link[i].link_identifier)
      != EOF) {
fscanf(file_pointer, "%d", &d_link[i].from.node_identifier);
fscanf(file_pointer, "%f", &d_link[i].from.node.x);
fscanf(file_pointer, "%f", &d_link[i].from.node.y);
fscanf(file_pointer, "%f", &d_link[i].to.node.x);
fscanf(file_pointer, "%f%*1c", &d_link[i].to.node.y);
i++; }
number_of_directed_links = i;
return (number_of_directed_links);
}
```

The area and interior area are not represented here. Algorithms for computing area and testing a point for inclusion within an area, or on the boundary are considered in Chapter 9. The polygon structure represents the simple polygon; the name is shortened since many applications deal only with simple polygons. This structure contains an identifier, a descriptor (such as the label "New York State"), and a single structure of type STRING to contain the boundary. The complex polygon is similar, except that the number of holes must be stored, and the STRING structure is multidimensional to store the holes. A maximum number of holes must be specified, stored in cart_obj.h as MAXHOLES. We could use C's linked list capability to store holes, that is, to store within the structure a pointer to a structure of type polygon, which constitutes a hole. While such a structure is elegant and allows complex polygons to contain any number

Function 5.20

```
/* Two-dimensional objects */
struct POLYGON { int polygon_identifier;
        char polygon_descriptor[MAXLABEL];
        struct STRING boundary; };
struct COMPLEX_POLYGON { int polygon_identifier;
        char *polygon_descriptor;
        struct STRING outer_boundary;
        int number_of_holes; struct STRING holes[MAXHOLE]; };
struct GRID { char grid_descriptor[MAXLABEL]; int nrows;
        int ncols;
        struct POINT corners[4]; int datum;
        int z[MAXROW][MAXCOL]; };
```

of holes without wasting storage, this addition will be left to the more advanced C programmer.

The pixel, being largely device dependent, is excluded from the list of structures. The reason for this is that the GKS standard provides a means for symbolizing a grid without reference to pixels. The remaining structure, then, is the grid. The grid has as its elements a grid descriptor, which could be the header lines from the USGS DEM data files, POINT references to the ground map coordinates of the four grid corners, which could, for example, be UTM values, and a two-dimensional array of integers to store the values. Elevations, on land range from -400 meters on the shore of the Dead Sea to 8,848 meters at the summit of Everest. In case the range of values required does not fit into the range of an integer, the datum value is provided. This integer value should be added to the grid value to give the actual elevation. For most DEM data, this is zero, but for ocean depths, elevations on other planets, and other geographic variables, other values may be necessary.

5.7 DATA STORAGE AND DATA STRUCTURES

At least one major problem with the data standards is the lack of three-dimensional cartographic objects. Fortunately, a small set of cartographic data structures for three-dimensional objects exists, which is covered in Chapter 6. In CADD, and in computer graphics generally, data structured for three-dimensional objects are a vital part of their symbolization or rendering as pictures. The GKS graphics standard is in the process of being extended to three-dimensional graphics, which may tend to favor particular three-dimensional data structures in the future.

The C language structures introduced in this chapter are used extensively in the chapters which follow. In Chapter 6, we use them to give examples of the data structures in common use, while in Chapters 9 to 11, we build upon them with transformational algorithms and symbolization methods using GKS. While their understanding is not critical to students of computer cartography, students of analytical cartography are strongly encouraged to use them to experiment with their own software for computer and analytical cartography.

5.8 REFERENCES

Defense Mapping Agency (1985) *Standard Linear Format*, Washington, DC.

Defense Mapping Agency (1985) *Feature Attribute Coding Standard*, Washington, DC.

Department of the Interior, United States Geological Survey, *Digital Line Graphs from 1:24,000-Scale Maps, Data Users Guide*, National Mapping Program Technical Instructions, Data Users Guide 1, Reston, VA, 1986.

Department of the Interior, United States Geological Survey, *Digital Line Graphs from 1:100,000-Scale Maps, Data Users Guide*, National Mapping Program Technical Instructions, Data Users Guide 2, Reston, VA, 1985.

Department of the Interior, United States Geological Survey, *Digital Line Graphs from 1:2,000,000-Scale Maps, Data Users Guide*, National Mapping Program Technical Instructions, Data Users Guide 3, Reston, VA, 1987.

Department of the Interior, United States Geological Survey, *Land Use and Land Cover Digital Data from 1:250,000- and 100,000-Scale Maps, Data Users Guide*, National Mapping Program Technical Instructions, Data Users Guide 4, Reston, VA, 1986.

Department of the Interior, United States Geological Survey, *Digital Elevation Models, Data Users Guide*, National Mapping Program Technical Instructions, Data Users Guide 5, Reston, VA, 1987.

Department of the Interior, United States Geological Survey, *Geographic Names Information System, Data Users Guide*, National Mapping Program Technical Instructions, Data Users Guide 6, Reston, VA, 1985.

HELD, G. (1983) *Data Compression: Techniques and Applications, Hardware and Software Considerations*, Wiley, New York.

LANGRAN, G. AND BUTTENFIELD, B. P. (1987) "Formatting geographic data to enhance manipulability", *Proceedings, AUTOCARTO 8*, Eighth International Symposium on Computer-Assisted Cartography, Baltimore, MD, March 29-April 3, pp. 201-210.

National Committee for Digital Cartographic Data Standards (1988) "The proposed standard for digital cartographic data", *The American Cartographer*, vol. 15, no. 1, pp. 9-142.

6

Map Data Structures

6.1 WHY MAP DATA STRUCTURES ARE DIFFERENT

A distinction can be made between map and non-map data structures. Map data structures are different because they store information about location, scale, dimension, and other geographic properties. Non-map data structures may be linked to the data structures for cartographic objects but they contain attribute information about the objects, or links between the objects only. In this chapter, we will consider only the data structures dealing explicitly with locative information. In the following chapter, we will focus on non-map data structures. Map data structures are particularly cartographic, and have been developed by cartographers and others to deal with computer and analytical needs, and for use in GISs. Non-map data structures can also be called data models, and are used to organize text or numerical data as well as cartographic data. Also essential is how these two types of data structures are related to each other, and how data can be transformed between structures. This theme is the topic of Chapter 10.

In the early days of computer cartography, the data sets which went along with cartographic software contained the minimum amount of information necessary to produce the desired map, and as such used data structures which were non-analytic, device specific, and as a result short-lived. These data structures, which have become known as entity-by-entity, or "cartographic spaghetti," remain in circulation, and for many purposes still serve the needs for which they were intended. Early on, however, attention became focused on the potential power of topologically encoded cartographic data. The group at Harvard, at the time producing the ODYSSEY package, sponsored a conference on topological data structures, which became a milestone in the acceptance of the need for topological encoding (Dutton, 1979). These topological structures have become the data structures for a large number of data sets such as DLG, GIRAS, and

TIGER, and form the core of a majority of the systems for the display and analysis of cartographic data.

In 1975, Tom Peucker (pronounced Poiker) and Nick Chrisman published a paper entitled "Cartographic Data Structures," which reported a way in which the information required for the topological encoding of cartographic data could be converted into a data structure which was particularly geographic. Their structure, part of a program called POLYVRT, became the model for many cartographic systems, and was highly influential in the evolution of data structures (Peucker and Chrisman, 1975). Since then numerous new, efficient, and thought-provoking data structures have been devised for cartographic data. The grid structure, the quad tree, and tessellations in addition to the entity-by-entity and the topological structures have found acceptance and use within computer mapping systems.

A *data structure* can be defined as a logical organization of information to preserve its integrity and facilitate use. By *information* here we mean the information content inherent in cartographic data, especially that necessary to produce a map. The information relates to the fundamental characteristics of geographic data. We have to *preserve the integrity* of this information. In other words, we have to store it with correct precision, in a coherent and consistent manner, and in such a way that we can retrieve the right information at the right time. To *facilitate the use* of the information we need to have some way of organizing our data physically on a storage medium in a method of representation which allows us to symbolize cartographic objects on maps with relative ease and within a reasonable length of time. Much work in computer cartography consists of writing data into cartographic data structures and transforming the data between structures. Different structures suit different kinds of mapping and different sets of demands.

Generally when we talk about cartographic data structures, the general characteristics are largely input determined. In other words, the data structures inherent in the geocoded data are usually determined by whatever input device was used to capture the data. Whether it was a digitizing tablet or an orbiting satellite, the structure that the data collection instrument imposed on the data is almost always the form in which we get the data. As the digital cartographic data standards gain acceptance, most existing data will be in more logical formats, but for the present, the input-determined nature of the data is a fact.

In many cartographic applications, the data need only be a simple representation of the map in order to produce a graphic. Integrating data sets, such as plotting the various separations which make up a multi-color topographic map, simply involves overlaying one data layer onto another. The cartographer need only decide upon an order of precedence. In additive color systems, the last color plotted at one location takes priority, and no mixing takes place, although there are exceptions. In these types of data sets we have no explicit relations between the cartographic objects. The only geographic property represented is location, and the map interpreter is left the task of searching for geographic relationships.

Thus we have two types of data structure in analytical and computer cartography. First, we have data structures to store strictly locational data about the map, with perhaps any necessary topological or other data required to produce the map. In addition, we have non-map data structures, which encode the attributes associated with cartographic objects, plus the indexing necessary to support query and retrieval functions on the objects. A map data structure is the minimum required for a computer mapping system, plus the actual data with its representational characteristics, and its origins in geocoding.

We can now respond to the question at the beginning of this section. Why are map data structures different? Map data structures are required for computer mapping, and they are different from non-map data structures in that their purpose is to support computer and not necessarily analytical cartography. A map plus a non-map data structure is the minimum requirement for the additional analytical functions we may use in analytical cartography, in the more sophisticated displays of advanced computer mapping systems, or in GISs. Also, many of these systems owe their power and capabilities to how the map and non-map data structures are related to each other, both logically and physically.

6.2 VECTOR AND RASTER TECHNOLOGIES AND DATA STRUCTURES

We have already seen a hardware division between raster and vector systems. Just as there are certain input controls on data structures, there are certain output constraints on data structures. Both input and display technologies have reflected generally a movement from vector-based to raster-based over time. Vector input devices such as digitizing tablets, and output devices such as pen plotters and storage tubes, are increasingly being replaced with raster technology, such as scanners for input and refresh-tube displays for output. This has influenced thinking about data structures considerably.

Both raster and vector data structures have been proposed as the answer for structuring geographic data. To use either one to the exclusion of the other, however, means accepting the inherent disadvantages of one of the systems. Many recent computer mapping systems and GISs have solved this problem by supporting both structures and allowing the user to transform between structures as appropriate. Peuquet (1979) showed that most algorithms using a vector data structure have an equivalent raster-based algorithm, in many cases more computationally efficient. Logically, therefore, the user should structure data according to the relative merits of each data structure for a particular type of cartographic entity, or a particular mapping circumstance. Burrough (1986) has provided an extensive list of the advantages and disadvantages of each type of data structure. The advantages and disadvantages are worth considering, since they reflect the correct choice of structure for dealing with particular types of data and particular geographic properties.

6.2.1 Advantages of Vector Data Structures

Burrough considered vectors a good representation of data mapped as lines and polygons, since vectors can follow the lines very closely. This means that maps produced from vector-based mapping systems are capable of greater resolution and accuracy. The vector structure is also compact, in that it can represent very complex lines with a minimal amount of information. We have seen that in vector mode we can reduce redundancy because if we have a fairly long line segment we only need to represent it with two x,y pairs, the two end points. Many other systems embed redundancy in representing straight lines.

As far as topological data structures are concerned, vectors are preferred, since topology can be completely described within vector-based systems as network linkages. Vector structures support interactive retrieval, so that updating and generalization of map data and non-map data attributes are possible. For example, if we have a vector representation of a line at 1:25,000, it is very easy to resample the points along that line to give a generalization of that line so that we can plot the map at another scale, such as 1:250,000. Also, in editing the map we can remove an entire chain and reenter it if we have made an error.

6.2.2 Disadvantages of Vector Data Structures

Vector data structures are complex and are less intuitively understood by the users of cartographic data. The biggest disadvantage, however, is that the combination by overlay of two or more vector-based maps is a very computationally intensive task. Such overlay, or computation of the greatest-common geographic units, means processing networks to find where they intersect. Just finding the intersections is difficult; since we have to test every line segment of every chain to find an intersection unless we use some kind of preprocessing based on the ranges in x and y of the chains. We next have to reorder the polygons by the new intersections. Overlay is a classic geographic problem, and many geographic phenomena need to be redistributed between sets of irregular areas. Not the least important, for example, is the redrawing of congressional and voting districts after each decennial census.

Display and plotting of cartographic data in vector mode can be expensive, particularly higher-quality color and cross-hatching. A vector-mode device especially does not do area fills very well, since plot devices try to fill areas by dragging the pen backward and forward, a time-consuming process. Vector mode display technology is also expensive, particularly for the more sophisticated software and hardware.

6.2.3 Advantages of Raster Data Structures

The principal advantage of raster data structures is their simplicity. The grid is an integral part of cartography and has long been used to structure geographic information. More recently, cartographic data have been acquired in grid formats directly, for example, satellite and scanned air photo data. This aspect is becoming increasingly important as remote sensing becomes a source of new cartographic data. Using the grid structure, analytical operations are easier, such as computation of variograms, some autocorrelation statistics, interpolation, and filtering. Resolution on raster devices is improving steadily, and the cost of the graphic memory boards which support display in raster mode is falling, whereas vector technology has not changed significantly in cost or capability in 10 years.

6.2.4 Disadvantages of Raster Data Structures

The principal drawback to raster data structures is the required volume of data. Grids embed much redundancy, so have a much larger data volume. If we have extremely variable data, such as topography, then grids use storage inefficiently, since compression methods do not save significant amounts of space. Also, the use of large cells to reduce data volumes means that some cartographic entities can be lost, simply slipping through the sampling net. Raster maps are considerably less visually effective than maps drawn with lines. Text on low-resolution raster devices is often illegible. Also, curved and angled text is difficult to produce. In raster mode, network linkages are difficult to establish. Cartographic entities such as stream networks, or boundaries between polygons, or linkages between town centers are not well supported as cartographic objects. It is difficult, for example, to establish links within a stream network in raster mode. Map projection transformations are time consuming using raster data structures, unless special algorithms or hardware are used. A disadvantage often ignored is orientation. If we use a grid, we imply an orientation to a grid, which automatically means that the sampling strategy has a particular directional bias. We normally orient our grids north/south and so bias the data against other directions.

As we have seen, neither structure holds all solutions for all data. Grids are usually used for mapping extensive geographic information at small scales, while vectors are used for detailed information at large scales. Systems which support both structures allow both.

6.3 ENTITY-BY-ENTITY DATA STRUCTURES

Cartographic entities are usually classified by dimension into point features, line features, and area features. The simplest means of digitally representing cartographic entities as objects is to use the feature itself as the lowest common denominator. Using entity-by-entity data structures, we are concerned with discrete sets of connected numbers which represent an object in its entirety, not as the combination of features of lesser dimension.

The simplest entity-by-entity structure is the point. In the digital cartographic data standards, the cartographic objects listed as "graphics only" are closest to entity-by-entity in type. Thus the point, especially the entity point, is the simplest form for dimension 0, the string, the arc, and the ring, for dimension 1, and the polygon and grid cell for dimension 2. Chapter 5 included program segments for reading and storing data in these structures, and Chapter 9 contains the necessary code for generating the map from the data in these structures. The grid cell structure is covered in detail later in this chapter.

An important difference between entity-by-entity and topological structures is shown by the ring structure. As far as a map is concerned, a lake can be shown on a map using a RING data storage structure. The fact that the first and last points match perfectly in x and y is simply to close the polygon visually. If we wish to compute the lake area, or the length of the boundary, or even shade the lake with color, the ring structure is adequate. However, if we wish to find the polygons which border the lake, or determine which rivers flow into the lake, it would require a large amount of systematic searching to find end-point matches. Similarly, for a point, we are able simply to plot a symbol at the point, but do not know which other features are related to the point and how.

6.3.1 Point Objects

Attributes associated with cartographic objects stored in entity-by-entity data structures rarely go beyond those contained in the C language structures in Chapter 5. Point features need only a feature code, because the feature code can be used as a key into a non-map attribute data-base of independent records for each feature. An entity-by-entity list of cities could contain their locations stored as a set of points, and if populations were required, for example, to produce a proportional circle map, the cities could be stored as label points and the label itself could be used as an index into a file of attributes. The following files could contain all the data necessary to produce a proportional symbols map.

Function 6.01

```
File : City_Populations
Format : ASCII population est. in 1986, source Goode's World Atlas,
         17th Ed.
Structure : Flat File

        Quito               918884
        Rabat               367620
        Rangoon             2276000
        Rawalpindi          452000
        Recife              1204738
        Riga                875000

File : City_Points
Format : ASCII Lat/Long in DDD.MM format
Structure : Label Point

        Quito               -0.17           -78.32
        Rabat               33.59           -6.47
        Rangoon             16.46           96.09
        Rawalpindi          33.40           73.10
        Recife              -8.09           -34.59
        Riga                57.56           23.05
```

6.3.2 Line Objects

In an entity-by-entity structure, a line is usually represented by a string, that is, an ordered set of points, which connected together in sequence trace out the line. In vector mode the string is unconstrained, although usually lines are terminated and restarted when crossing takes place. Similarly, there is no obstacle to double specification of a single line, a common problem resulting from the manual digitizing process. A line can carry an attribute, such as a traffic volume, a color, a thickness, or a feature type such as highway or railroad. Attributes are not really assignable to strings directly, but a separate file of attributes by string number can be maintained. Alternatively, separate files for each feature type can be maintained, similar to separating line colors by separation in a manual color production process.

A common alternative to keeping a file containing each string listed point by point is to use a point dictionary (Figure 6.01). In this file arrangement, separate files are kept for points and for lines. The point file is simply a listing of every point used on the map, containing an identifier, often sequential, and an easting and northing. The second file, the line file, contains line numbers, perhaps the line attributes, and the starting and ending point numbers of sequential groups of points in the points file which make up the line. The point numbers act as pointers into the point dictionary. An alternative system has a list in the lines file of every point identifier in the line. This system, while much more cumbersome to implement and use, has the advantage that the point data file can be sorted at will to facilitate searching or processing. Point dictionary systems make editing and data entry relatively difficult, but have found their uses in computer mapping systems.

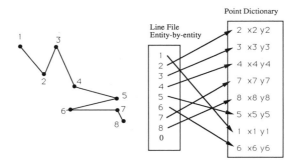

Figure 6.01   Point dictionary files

Several grid-type structures exist for entity-by-entity data structures for lines. If a grid is used, lines are typically labeled with attribute numbers which are assigned to grid cells as data. Thus to give an entity-by-entity definition of a river for use in a grid mapping system, we could assign a value for the river to every pixel into which the river passes. Connectivity can be limited to the four adjacent cells or enlarged to permit diagonal or eight-cell connectivity. It is normal to thin the lines to one pixel width for

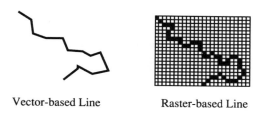

Vector-based Line   Raster-based Line

Figure 6.02   Presentation of lines in entity-by-entity structures

processing and display (Figure 6.02). Single lines in one map are often stored as bit planes, with run-length encoding. If lines are stored as attributes or index numbers in the grid, problems result at line intersections. These are sometimes stored separately; otherwise, the attribute or color plotted at an intersection is the last one referenced.

The Freeman code is an entity-by-entity line data structure which falls between vector and grid. The Freeman code is a representation of a line as a sequence of numerical codes, each representing the direction of a step moved along the line (Figure 6.03). Freeman codes which use one octal digit per code are eight-pointed, while one hexadecimal digit (nibble) per code allows 16 values. A problem of converting grid data to Freeman codes is that diagonals have length of the square root of 2 over 2 while the primary directions have length $1/2$ in grid units. Both strategies can be used. Anstaett and Moellering (1985) gave a simple algorithm for computing areas of polygons bounded by sets of Freeman codes. Freeman codes can also be run-length encoded, meaning that files containing long straight lines, represented as repeating identical Freeman codes, can be compressed considerably.

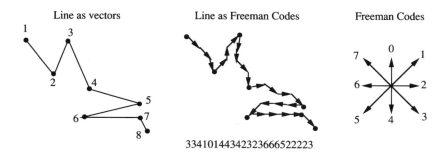

Line as vectors   Line as Freeman Codes   Freeman Codes

33410144342323666522223

Figure 6.03   Freeman codes as a line data structure

6.3.3 Area Objects

Since we are dealing with entity-by-entity structures, polygons of two-dimensions can be represented cartographically by one-dimensional objects unless we wish to fill the polygon. Appropriate structures in the C language, as given in Chapter 5 are the RING, in its form as a set of strings or as a set of arcs. As a two-dimensional object, the POLYGON structure is appropriate. Strictly speaking, the complex polygon with holes cannot be stored as an entity-by-entity object. We can include the hole as part of the boundary, with a "bridge" between the two lines, or we can use alternative means to include the hole.

Just as we used the point dictionary for a line above, we can also use the point dictionary for an entity-by-entity area (Figure 6.04). In this case, the second file is the area file, and contains the polygon identifier and a list of the points which make up the boundary. There may be a good reason for using this structure, since a sorting of the point dictionary makes it simple to determine shading instructions for filling the polygon. A polygon list, in addition, is the simplest means for computing the area of a polygon and for drawing the polygon as a single entity against a plain background on a map.

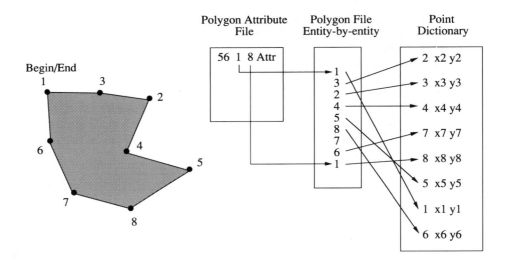

Figure 6.04   Point dictionaries for a polygon

Using the grid structure, polygon representation is simple. Membership is usually assigned to the grid cell with the highest area content for any polygon, and all cells within a polygon are assigned one value, either an index or an attribute (Figure 6.05). Alternatively, the class of the center of the grid cell can be assigned. Holes can simply be assigned last, so that their attributes fall on top of any prior values. In this way, areas can be computed by cell counting, a rapid though less accurate technique. Similarly, polygons can be made up of strings of Freeman codes, though in this case care must be taken to assure closure. Closure is when the vector sum of the individual Free-

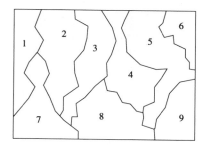

```
1111122222223333334444455555555666666
1111222222222223333334455555555566666666
1111112222222223333334455555555556666
1111122222222223333334445555555555566
1117772222223333333444444455555555556
1177777222223333333444444444444555555
1111772222222333388444444444455555599
1117777223333388888884444444455555999
1777777722333888888888888445559999999
7777777777338888888888888885599999999
7777777777777888888888888888899999999
```

Figure 6.05   Polygon structures for a grid

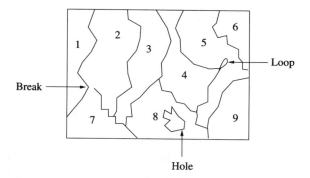

Figure 6.06   Weird polygons

man code vectors is zero, but again no requirement to prevent crossing is included. Crossed-over polygons (Figure 6.06) are sometimes called weird or splintered polygons. Area calculations and other operations are not appropriate for these polygons.

6.4 TOPOLOGICAL DATA STRUCTURES

Entity-by-entity data structures, while they are useful for the symbolization of cartographic objects, are often unsuitable for more complex analysis, since they store no information about topology. Furthermore, we have no guarantee with entity-by-entity structures that impossible geometric configurations such as weird polygons and slivers do not exist. Topological data structures store the additional characteristics of connectivity and adjacency. As such, they allow immediate checking, perhaps during data entry, for errors in geocoding and data storage. The basic objects are the node, the link, and the chain. To these are added direction, as in the directed link, and higher structure, as in the network chain. Nodes are points with topological significance, and as such must be labeled. The directed link and the network chain are similar in that they support linkages to adjacent objects and nodes. Connectivity to nodes is essential to give a graphic frame of reference and to link lines in the network together. The direction, and the links to other primitive objects, can be summarized as properties of a winged segment (Figure 6.07).

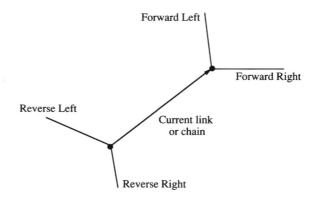

Figure 6.07   A winged segment

    Linkages are sometimes stored as forward linkages and reverse linkages. This is because chains can begin only at one end, and to piece together objects it is very often required to traverse a chain in either direction. Forward linkages are the segment identifiers of connected segments pointed to when a chain is traversed in the direction in which its x and y values are stored, We can have forward linkages to other links, directed links, or chains. Usually, there is a finite limit to the number of chains which can meet at a node. This is not a problem for most maps; for example, the 48 contiguous states have only one node where four lines meet (four corners), but sometimes, for example at the poles, many lines meet together at once.

    To traverse an entire network we only need to know immediate right- and left-hand turns. Eventually, if we keep following around the network, you will come back to the start point along some other chain. If we have made an error and one of the chains has not been snapped to a node, or if not all intersections are nodes, the network tracking algorithm in trying to close the polygon will go off along the wrong chains, going around through the network trying to close a polygon.

    Topological data structures also store polygon information, which makes the data structure more analytically powerful. We assign a right-hand polygon and a left-hand polygon according to the direction of the chain or directed link. This allows us to presort the chains, for example to select all chains that say that they are neighbors of a particular polygon. These chains can then be connected either by their chain linkage information or by matching their end points, and converted into a polygon list as an entity-by-entity structure or simply plotted in order.

6.5 TESSELLATIONS AND THE TRIANGULATED IRREGULAR NETWORK

Tessellations are connected networks which partition space into a set of sub-areas. They consist of the network itself and the spaces left within the areas. Special cases of tessellations are topological structures, in which the space is partitioned into regions of geographic interest such as states and counties; and grids, in which the tessellation is a network of connected squares and the grid cells they divide the space into. The remainder of tessellations include a number of systems used infrequently in analytical and computer cartography, such as equilateral triangles and hexagons, which are geometric and regular in nature, and one irregular system which has gained widespread use, the *triangulated irregular network* (TIN). Mark (1975) has suggested that TINs are more accurate and use less space than any other data structure for topography, and McCullagh and Ross (1980) demonstrated that TINs can be generated from point data faster than the alternative data structure for terrain, the grid. TIN data structures can describe more complex surfaces than a grid, including vertical drops and irregular boundaries. Since single points can be easily added, deleted, or moved, this structure has gained widespread acceptance in CAD systems, surveying, engineering, and in terrain analysis.

The TIN structure was proposed by Peucker et al. (1976), and takes into account an important property of geographic data collection, i.e., that map data collection often tabulates data at points, and that the points themselves have an inherent significance as far as the information content is concerned. For example, a surveyor measuring land surface elevations seeks "high information content" points on the landscape, such as mountain peaks, the bottoms of valleys and depressions, and saddle points and break points in slopes, and measures the elevation at these points. The intervening elevations are then often interpolated from the significant points. The simplest form of interpolation is to assume that between triplets of points the land surface forms a plane. This is the foundation of the TIN structure. Triplets of points forming irregular triangles are connected to form a network. Rules are followed in partitioning the points into the nodes of the triangles. The space is then divided into a set of triangles, their edges, and their nodes. Data can be associated with any of these elements. Two approaches are in common use to arrive at the "best" triangulation. In the first method, an initial triangulation is assigned, and is then refined until the triangles meet the Delaunay criterion. Alternatively, some methods seek to compute the best triangulation in one step. McKenna (1987) reviewed these approaches.

The first problem in TIN creation is how to allocate the triangles between points. This problem is usually solved using the Delaunay triangulation (Figure 6.08). This is an iterative solution, which begins by searching for the closest two nodes, and then assigns additional nodes to the network if the triangles they create satisfy a criterion such as selecting the next triangle which is closest to a regular equilateral triangle. Many different methods for performing this space partitioning exist and are constantly being improved and made more efficient (McKenna, 1987).

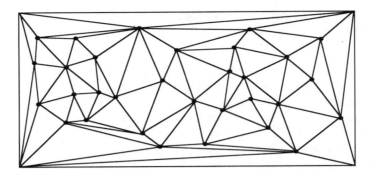

Figure 6.08   Delaunay triangulation to yield a TIN

The Delaunay triangulation yields a convex hull (Figure 6.09). This is because if we make an attachment which makes a concavity in the edge of the network, the network can be extended until the concavity is removed. So in other words, there are no concavities in the hull's edges. The convex hull presents another problem in constructing a TIN, because no data can be stored in the structure for the region between the edge of the convex hull and the square or rectangular edges of the map. A common solution is to include the map corners, and perhaps points along the map edges, in the Delaunay triangulation.

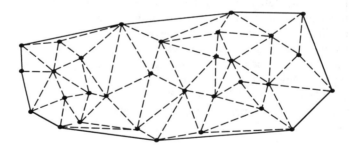

Figure 6.09   A convex hull

Two data structures have been used to physically represent TINs within the computer's memory. The first uses the triangle as the basic cartographic object, while the other uses the vertices of the triangles. Most TIN structures use the triangles as the object, and store links to neighboring triangles (e.g., Gold et al., 1977). In this case, data are contained in two files (Figure 6.10). The first file contains points, stored as [x,y,z] triplets with their eastings, northings, and elevations. The second file contains one record per triangle and contains three attributes which are pointers into the point file, plus three additional pointers to adjacent triangles. The following code sample gives an example of data stored in this data structure.

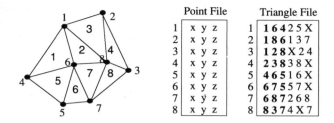

Point File		Triangle File	
1	x y z	1	1 6 4 2 5 X
2	x y z	2	1 8 6 1 3 7
3	x y z	3	1 2 8 X 2 4
4	x y z	4	2 3 8 3 8 X
5	x y z	5	4 6 5 1 6 X
6	x y z	6	6 7 5 5 7 X
7	x y z	7	6 8 7 2 6 8
8	x y z	8	8 3 7 4 X 7

Figure 6.10   The TIN data structure based on triangles

Function 6.02

```
/*    ----------------------------------------------------------
/*    Sample program to demonstrate the TIN structure using
/*    the triangle as the basic object : kcc 3-89
/*    ----------------------------------------------------------
*/
    #include "cart_obj.h"
    #define MAXTRIANGLES 200
    void tin_info_triangle () {
    struct TRIPLET { struct POINT point; int elevation };
    struct TIN { int vertex [3], int neighbors[3] };
    struct TIN tin[MAXTRIANGLES];
    struct TRIPLET triplet[MAXPOINTS];
    int i;
    /* Here we assume that the TIN is read from a file */
    read_tin_triangle();
    /* Here we assume that the points are read from a file */
    /* Print out the data for triangle 16  (index 15) */
    printf("Triangle 16 Data\n");
    for (i=0;i < 3; i++)
    printf("Node %d at (%f,%f) has elevation %d\n",
         i, triplet[tin[15].vertex[i]].point.x,
         triplet[tin[15].vertex[i]].point.y,
             triplet[tin[15].vertex[i]].elevation);
    printf("Adjacent triangles are :\n",
    for (i=0;i < 3; i++)
        printf("Triangle number %d\n",tin[15].neighbors[i]);
    }
```

The original TIN structure used in Peucker and Chrisman's paper (Peucker and Chrisman, 1975), used a vertex-based system. In this data structure, the point record contains [x,y,z] as before, but also a pointer to a "connected points" file. The connected points

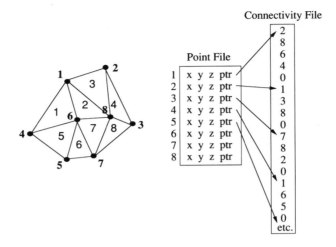

Figure 6.11   The TIN data structure based on vertices

file is sequential and contains at each location pointed to in the points file a list of nodes which are connected to the point in question. The list is terminated with a zero to allow an arbitrary number of nodes (Figure 6.11). This data structure uses about half the storage of the triangle object data structure. Each of these two structures has advantages and disadvantages depending on the particular application. In both cases, the second file is constructed from the points file during the Delaunay triangulation phase of the TIN construction.

A major advantage of the TIN structure is that many of the problems associated with automated contouring, hidden-line processing, and surface shading for cartographic symbolization are simpler to tackle in this structure. The structure is particularly suitable for workstations which support polygon manipulation and shading in three dimensions in hardware, allowing interactive production of perspective views and realistic perspectives which would take hours of computing time to generate using the grid data structure. The structure is also advantageous for modeling, especially of hillslopes and streams, which flow along the edges of the triangles across the terrain. The TIN data structure is also very suitable for the intervisibility problem, which seeks to determine in three dimensions solutions to which facets in the TIN are visible from points and lines, perhaps at different altitudes.

6.6 QUAD TREE DATA STRUCTURES

The quad tree is a data structure which has received increasing use in recent years. At first, quad trees were used exclusively in image processing for binary images. More recently, a number of advances in algorithms have made quad trees a viable data structure for cartographic data. Quad-tree equivalent algorithms now exist for area computation, centroid calculation, image comparison, connected component labeling, neighbor detection, distance transformations, regionalization, smoothing, and edge enhancement (Tobler and Chen, 1986). Mark and Lauzon (1985), among others, pointed out the feasibility and advantages of the data structure for map data, while Samet (1984) provided the definitive survey.

Quad trees are areal indexing systems, and assume that the basic data element is the area, as represented by the extent of a grid cell. Quad trees, used as a cartographic data structure, also assume that all areas to be stored are grid cells with a side which is some power of 2 of the smallest resolution element. The quad tree data structure allows very rapid area searches and relatively fast display. The gain in efficiency is the result of the exploitation of the fact that areas on maps are typically some combination of large homogeneous areas and smaller, heterogeneous areas. Quad trees are tessellation data structures, in that the space on the map is partitioned by a covering network of spaces, but for the quad tree, the partitioning is into nested squares. Division of squares is into quadrants, giving the "quad" part of the name.

At the highest level, the entire map can be thought of as one square. Rectangular or irregular maps can be subdivided into a set of square regions to become starting points in the quad tree. Each square is then divided into four quadrants, the NE, NW, SE, and SW, if and only if the square contains detail at a level greater than the current area of the square. The path to each quadrant then becomes a branch, from the root of the quad tree to its leaves, which are the highest level of detail. This system is familiar to users of the United States Public Land Survey System. In this system, square regions (and rectangular, and indeed irregular shapes, due to errors) are given a reference within a township and range system. For example, a small land tract may have the reference

Northeast Quarter of the Southwest Quarter of the Southeast Quarter of
Section 21, Township 24N, Range 2E.

When the land holding is large, it can be referenced simply; for example, we could record that all of Section 21, Township 24N, Range 2E is farmland. When the landholdings are small and various, we need the full list to specify the plot.

To index a very small area within a quad tree structure, we could use a list such as NE, SW, NE, NW, SE, etc. More efficient, however, is to store the Morton number. The Morton number is a unique identifier assigned to quad tree divisions as they are constructed. Figure 6.12 shows how Morton numbers are assigned. In their simplest form, quad trees consist of strings of Morton numbers, each of which contains the index or pointer to a data value associated with the highest-level quadrant. When large squares are sufficient, the Morton sequence is short, and vice versa.

Quad trees are most efficient when the basic cartographic entity is the polygon. The cells in the quad tree then become the cartographic objects, and the data structures contain the coordinates of the region in question, the Morton sequences for each unique cell division in the quad tree, and the attributes for these cells. More than this, the effi-

000	001	010	011	100	101	110	111
002	003	012	013	102	103	112	113
020	021	030	031	120	121	130	131
022	023	032	033	122	123	132	133
200	201	210	211	300	301	310	311
202	203	212	213	302	303	312	313
220	221	230	231	320	321	330	331
222	223	232	233	322	323	332	333

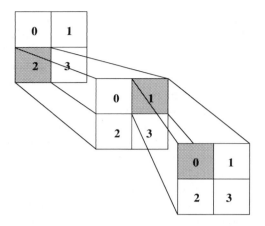

Morton Sequence 210

Figure 6.12   Morton numbers for quad-tree data structures

ciency of the quad tree is at its highest when a mix of large homogeneous areas and smaller polygons exists, since the storage gain is mostly from two-dimensional run-length encoding of the large regions.

Land use maps are good examples, when this is the case, as also are political units such as counties. Land use maps generally have a background category which occupies most of the space, for example forest or agriculture. The map then has patches on top of this background, roads, commercial districts, or lakes and wetlands. In these cases, the quad tree may be the most effective structure for several types of analysis to be performed on the data. Given that many applications in remote sensing generate regionalized or "segmented" grid maps, the strong link among the quad tree, remote sensing, and the grid data structure is self-evident.

6.7 MAPS AS MATRICES

The logical organization of raster data is easily converted into a physical representation in storage. This is because almost all computer programming languages support the array as a data structure directly. Thus in C, the array is given as z[i][j], in Pascal as z[i,j], and in FORTRAN as Z(I,J) As such, the mathematical term for this organization is the *matrix*. In simple terms, we can consider a map to be a matrix. Each element in the matrix falls as a grid cell on the ground, and can be set to coincide with the spacing and orientation of the coordinate system and map projection in use. For example, we could divide the world up into grid cells based on 10-degree increments of latitude and longitude, and store data for the cells in a matrix. Given the corners of the matrix, and the spacing, we can implicitly refer to each matrix element just by referring to its row and column numbers, rather than the explicit way we give eastings and northings for points (Figure 6.13).

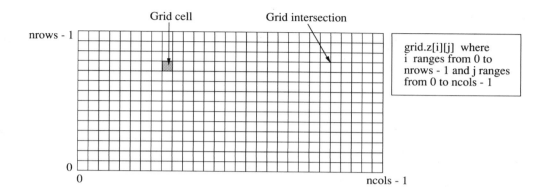

Figure 6.13   Generic structure for a grid

When we read a matrix into the computer's memory, we read it row by row and column by column, one element at a time. Having the grid stored in this way has the distinct disadvantage of simplicity. The disadvantage is that elements that are spatially next to each other are not necessarily close together in the file. Fortunately, when the data are within the array in a program, the immediate neighbors can be found easily.

Most automated mapping systems that use pixel-based display or use remotely sensed data in any form use the grid as their basic data structure. Usually, the actual files stored contain the registration information, such as the coordinates of the corners, the grid spacing, etc., plus the image itself, often in binary or compressed form to save storage. Files can be very large using the grid structure. For example, many display devices support images of 512 rows by 512 columns. A single data set, storing one byte per record, takes up 262 kilobytes without compression.

Both the digital cartographic data standards and GKS define a basic grid as a primitive element. Under the data standards, the elements are the grid cell and the pixel. Under GKS, the elements are the cell array and the pixel. As in the data standards, the GKS cell array grid cell need not be a single pixel. Using these elements in GKS is termed raster graphics, defined as a display image composed of an array of pixels arranged in rows and columns. Since this terminology is part of the standard, the issue of raster or vector representation is trivial, and either type can be displayed on any device.

The following function reads a grid into an array inside a C program. Code to generate a map using these data is given in Chapter 9. Appendix C includes a listing of a program to read USGS DEM data tapes into a grid format.

Function 6.03

```
/* _____
/*  Read data in ASCII integer format into a grid structure
/*  kcc     2-89
/* _____
*/
#include "cart_obj.h"
void read_grid ()
     struct GRID grid;
{
   FILE    *infile;
   char filename[30];
/* Prompt for data file information */
   printf ("\n\n Enter name of file containing data :");
   scanf ("%s%*1c", filename);
   printf ("\n Data are to be read from file: ");
   printf ("%s", filename);
/* Open the data input file */
   if ((infile = fopen (filename, "r")) == NULL) {
   printf ("\n Error: Cannot open your input file \n");
   exit ();
   }
     printf ("\n Enter a description of the grid :");
   printf ("%s%*1c", grid.grid_descriptor);
     printf ("\n Enter the number of rows    :");
     scanf ("%d%*1c", &grid.nrows);
     printf ("\n Enter the number of columns :");
     scanf ("%d%*1c", &grid.ncols);
     printf ("\n Enter the lower left and upper right x values :");
     scanf ("%d%d%*1c", &grid.corners.x[0], &grid.corners.x[2]);
     grid.corners.x[1] = grid.corners.x[0];
     grid.corners.x[3] = grid.corners.x[2];
     printf ("\n Enter the lower left and upper right y values :");
     scanf ("%d%d%*1c", &grid.corners.y[0], &grid.corners.y[1]);
     grid.corners.x[2] = grid.corners.x[1];
     grid.corners.x[3] = grid.corners.x[0];
     printf ("\n Enter the datum [default of zero] :");
     if (scanf("%d%*1c", &grid.datum) == NULL) grid.datum = 0;
   printf ("\n\n Data input in progress\n");
       for (i = (nrows - 1); i >= 0; i--) {
           for (j = 0; j < ncols; j++) {
             fscanf (infile, "%d", &temp_int);
             grid.z[i][j] = temp_int;
           }
       }
   return;
   }
```

Examples of data stored in a grid format have been given in Chapter 5, and include the DEM. Issues associated with grid data are registration problems, map projections, boundary effects, and edge matching, especially when the map edges do not neatly fit along the edges of the grid.

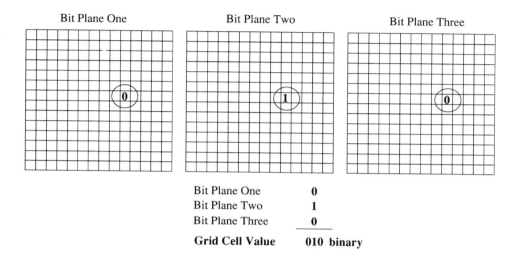

Bit Plane One	**0**
Bit Plane Two	**1**
Bit Plane Three	**0**
Grid Cell Value	**010 binary**

Figure 6.14   Bit plane storage of a grid map

An alternative way of storing information in an array is to use bit planes (Figure 6.14). Bit planes are much more associated with storage and display than with manipulation, and are particularly useful for large areas with similar values, although they are universally applicable. In some cases, bit-mapped terminals can be used directly using this structure. Very high resolution monochrome terminals are often bit-mapped.

Bit planes can be used in two ways. Bit planes can represent colors on a display as a map of the pixels. One bit plane can be made to correspond to each color. For the six major colors (red, green, blue, magenta, cyan, and yellow), we need only three bit planes. For a single cell, if all three planes are 0, the screen shows black. If all three planes are 1, the screen shows white. Red, green, and blue correspond to the three planes directly, with a 1 in the correct plane determining the color. Magenta has a 1 in red and blue, cyan has a 1 in the green and blue bit planes, and yellow has a 1 in green and red. Thus with three bit planes, we can store eight colors, and since the bit arrays can be placed directly into the video channels in the display, a color screen can be drawn quickly. Four and more bit planes become complex to use, although storing data bit by bit in bit planes allows extremely fast overlay of images and permits binary operations between planes, such as AND, OR, and NOR. Each bit plane, in addition, can be run-length encoded to drastically reduce the amount of memory required (Figure 6.15). In run-length encoding, we store pairs of bytes, the first representing the value stored in the array, the second giving the number of times that the value should be repeated. Very large space savings are possible using run-length encoding when much of the map is homogeneous.

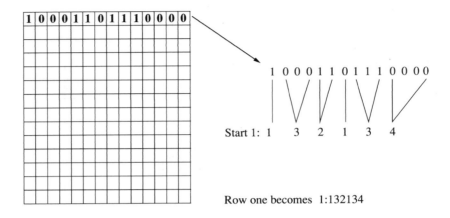

Figure 6.15　Run-length encoding for a bit plane

6.8 MAP AND NON-MAP DATA STRUCTURES

In this chapter, we have surveyed the major data structures in use for the storage of cartographic objects within their mapping context. Analytical and computer cartography, however, requires far more than simply the map data to produce an effective map, to use the data for new cartographic applications, and to support the demands of GISs. In Chapter 7 we will consider how non-map data can be structured, and examine the need for effective links between the map and the non-map data. It is on the strength of the map-to-non-map link that many mapping systems and GISs will stand or fall as new and innovative uses for cartographic data find their way into the everyday life of non-cartographers. We should remember, therefore, that a good cartographic data structure must both preserve the data integrity and make the geographic information it contains available to make maps which can be used and understood by all.

6.9 REFERENCES

ANSTAETT, M. R., AND MOELLERING, H. (1985) "Area calculations using Pick's theorem on Freeman-encoded polygons in cartographic systems," *Proceedings, AUTOCARTO 7*, Seventh International Symposium on Computer-Assisted Cartography, Washington, DC, March 11-14, pp. 11-21.

BURROUGH, P. A. (1986) *Principles of Geographical Information Systems for Land Resources Assessment*, Monographs on Soil and Resources Survey Number 12, Oxford Science Publications, Clarendon Press, Oxford.

DUTTON, G. (Ed.) (1979) *Harvard Papers on Geographic Information Systems*, First International Advanced Study Symposium on Topological Data Structures for Geographic Information Systems, Addison-Wesley, Reading, MA.

GOLD, C. M., CHARTERS, T. D., AND RAMSDEN, J. (1977) "Automated contour mapping using triangular element data structures and an interpolant over each irregular triangular domain," *Computer Graphics*, vol. 11, pp. 170-175.

MARK, D. M. (1975) "Computer analysis of topography: a comparison of terrain storage methods," *Geografiska Annaler*, vol. 57, pp. 179-188.

MARK, D. M. AND LAUZON, J. P. (1985) "Approaches for quadtree-based geographic information systems at continental or global scales," *Proceedings, AUTOCARTO 7*, Seventh International Symposium on Computer-Assisted Cartography, Washington, DC, March 11-14, pp. 355-364.

MCCULLAGH, M. J. AND ROSS, C. G. (1980) "Delaunay triangulation of a random data set for isarithmic mapping," *Cartographic Journal*, vol. 17, no. 2, pp. 93-99.

MCKENNA, D. G. (1987) "The inward spiral method: an improved TIN generation technique and data structure for land planning applications," *Proceedings, AUTOCARTO 8*, Eighth International Symposium on Computer-Assisted Cartography, Baltimore, MD, March 29-April 3, pp. 670-679.

PEUCKER, T. K., FOWLER, R. J., LITTLE, J. J., AND MARK, D. M. (1976) *Digital Representation of Three-dimensional Surfaces by Triangulated Irregular Networks (TIN)*, Technical Report Number 10, United States Office of Naval Research, Geography Programs.

PEUCKER, T. K., AND CHRISMAN, N. (1975) "Cartographic data structures," *The American Cartographer*, vol. 2, no. 1, pp. 55-69.

PEUQUET, D. J. (1979) "Raster processing: an alternative approach to automated cartographic data handling," *The American Cartographer*, vol. 6, pp. 129-139.

SAMET, H. (1984) "The quadtree and related hierarchical data structures," *IEEE Transactions on Pattern Analysis and Machine Intelligence*, vol. PAMI-4, no. 3, pp. 298-303.

TOBLER, W. R. AND CHEN, Z. T. (1986) "A quadtree for global information storage," *Geographical Analysis*, vol. 18, no. 4, pp. 360-371.

7

Non-Map Data Structures

7.1 SEQUENCE OF DEVELOPMENT

In the preceding chapters we have examined physical data structures, the actual arrangement of digital cartographic objects in computer files. We have also examined logical data structures, the conceptual links between digital cartographic objects and their physical data structures. Since the advent of geographic information systems (GISs), much attention has been devoted to how effective the physical-to-logical data structure link is, especially when using these structures for interactive query and analysis. These additional functions, additional, that is, to computer cartographic systems, require the ability to query thematic data using geographic properties. More strictly we query the cartographic objects representing the geographic phenomena by their attributes reflecting geographic properties. In addition, these queries require the retrieval of data from many different storage locations, usually many files with different sizes and formats.

Advanced analytical operations, the sort supported in GISs, require the management of non-map data to support query and analysis by cartographic data structures. The map and the non-map data physically reside in separate files, or are separated within the computer's memory. The problem of managing these files is one of data-base management, a field well represented within computer science. If we look at the sequence of how data-base management systems developed, we find that there are distinctive trends in the historic development of non-map data structures. While these trends are more important to GISs than to computer cartography, they are of relevance to analytical cartography, and therefore will be discussed here.

The first generation of GISs and computer cartographic systems did not really support query and analysis interactively. These early systems had sophisticated map data structures, but were usually grid or entity-by-entity based, and often included the thematic data as part of the cartographic data structure. Since the purpose of these systems was to produce a map, the independence of the data was sacrificed. For example, to convert choropleth data from numbers to percentages would usually involve reentering the thematic data, perhaps as a set of parameters on "cards" for batch processing. The multi-file problem was avoided by merging the data into a single file containing the data structure.

Second-generation systems were still entity by entity, but they were topological systems. Using these systems, we can expand the basic entity of a point to build lines, and then lines to build polygons. Polygons in a topological system could consist of a chain list or a point list, but we also have as part of the data structure information about which chains are linked to which other chains. By keeping track of forward and reverse linkages so that we can reassemble our polygons if we need them to fill them or verify them, we are actually storing non-map data as raw geographic properties, in particular connectivity and contiguity. If we need to know which lines are connected to other lines, we can find that out from our data structure.

In these systems, different parts of the structure could be stored in different files. As a result, the data structure is more flexible. The thematic data, however, often remained as part of the map information. For example, the DIME files consisted of single sets of records, with index numbers for the polygons. The user of the DIME files had to manually merge the census variables with the map data to produce computer-generated maps. The only physical link between the topological and the non-map data was the census tract or block number, which happened to be the same in each of two independent data sets. While this arrangement was closer to supporting query and analysis, the data structure itself was not the means by which query was performed.

Third-generation systems combine map and non-map data together, usually within a GIS. Third-generation systems are distinctive because they are capable of performing separate data management, that is, performing transformations on the thematic data directly without reference to the geography and across multiple files. These systems have been able to integrate the capabilities of data-base management systems, such as selective retrieval, fast sorting, and operations on multiple data attributes, into systems designed for the display and analysis of geographic data. These systems use the concept of the data model, and implement one of three basic models common in data-base theory.

The first two of these data models, the hierarchical and network models, are considered in the following section. The third of these models, the relational model, is itself a movement to a fourth stage of non-map data structure development. As we will see, it is this data model which has allowed the development of GISs capable of performing real-time query and analysis by geographic properties and their pertinent cartographic objects. It is, therefore, worth examining the basic data models before returning to an examination of some current non-map data structures found in contemporary GISs.

7.2 NON-MAP DATA MODELS AND DATA MANAGEMENT

7.2.1 The Hierarchical Data Model

In the management of non-map data, the equivalent of the term *cartographic data structure* is the term *data model*. Data models were developed from data-base management theory, part of computer science. The first and simplest of the data models is the hierarchical model. The hierarchical way to structure data can refer to many different attributes, not just those of geographic data. Hierarchical data management systems form the basis of many commercially available systems, and have influenced file structures and our thinking about organizing data generally.

To use a geographical example, though, the top of a hierarchy is called the root (from the root of a tree) and for geography the root could be the WORLD (Figure 7.01). We know that there are seven CONTINENTS, and one of those continents is N. AMERICA, which consists of several countries, one of which is the United States of America. The USA consists of 50 states, one of which is New York, and New York consists of counties, one of which is New York (same name, but different level in the hierarchy). New York County, otherwise known as Manhattan, contains COMMUNITIES, one of which is the Upper East Side, where we can find Hunter College.

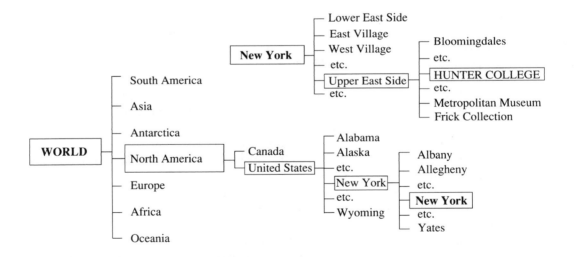

Figure 7.01   A geographic hierarchy

This is a hierarchical structure. Each piece of the structure has links to children and a single parent through which all searching must pass. A full reference to a particular element, such as a cartographic entity like Hunter College, a unit upon which we are going to hang data, gives its path down the tree. An atlas is usually structured this way, with all maps for Asia bound together and separated country by country. At the lowest level of the tree, if we think of the tree as inverted, we find the leaves or records. For cartographic data the records are most commonly cartographic objects, and we assume that the map data-base contains graphic primitives associated with the cartographic objects. In addition, normally for each entity we have attribute information. Attributes can store nominal, ordinal, interval, or ratio data, as well as text or complex characteristics.

Attributes and entities have both a physical and a logical structure. In data-base management systems, the physical structure is often hidden. Logically, the attributes and entities can be thought of as occupying a flat file. A flat file (Figure 7.02) resembles a spread-sheet table or matrix, with rows of records and columns of attributes. A flat file corresponds to a particular level of the hierarchy. For example, a flat file for New York State could contain counties as records, with each county having another flat file further down the hierarchy. The census distribution tapes and the city and county data book are good examples of flat files. Usually, all the communities are listed as rows, and hundreds and hundreds of data values for different attributes are listed as columns.

Attributes

State	Population	Area	% Water	Etc
Alabama	———	etc	etc	etc
Michigan	———			
New York	———			
———	———			

(Entities)

Figure 7.02   A flat file

In the hierarchical data model, we organize the links between the logical and the physical aspects of the data in a tree structure. While this system is often adequate, several problems can result. A major problem is cartographic objects which belong to multiple hierarchies. For example, police districts, school districts, census tracts, voting districts, and congressional districts rarely coincide, all being part of different hierarchies. Yet a single cartographic entity falls into all of these areas. Also, hierarchies are rarely spatially mutually exclusive, leaving gaps in some areas and double covering other areas. For example, the boundary between New York and New Jersey is marked at the half-way point on the George Washington Bridge. If the bridge is a single cartographic object in a cartographic data base, which state is it included under? Also, while cities are usually parts of counties, New York City contains five counties, an inversion of the usual direction of the hierarchy.

A simple geographic type of query of a hierarchically structured non-map database might be: How many people live in Manhattan community district number 4? Does it cross any census tract boundaries? An example of hierarchical cartographic information is contour lines on a map. The 100-meter contour line presumably contains all points with elevations greater than 100 meters, unless annotated as a closed depression. Hierarchical space partitioning may therefore be suitable for map data but unsuitable for all but the most rigidly encoded non-map data. The method also involves duplication and therefore is not highly storage efficient.

7.2.2 The Network Data Model

The second data model which has become common in data-base management theory is the network model. In the network model, we allow relationships between entities. Normally, these are not complex relationships. For example, we might allow an entity, a particular block in Manhattan, to belong to a certain congressional district, a census tract, and a school district. Here, the entity is a block and the relationship is a link to an area.

In computer science, a particularly powerful data structure is the linked list. In a linked list, each member of a list contains a pointer to the next member of the list. We saw in Chapter 5, for example, in the USGS's GIRAS files, that the polygon files simply contain sequential pointers to records in the arc files, which in turn contain pointers to the coordinate files. This is the linked list concept, and this concept is behind the network structure. Links between objects in this structure form networks. Geographically, the pointers carry associations such as "... is a member of". Using the parent-child analog above, network systems allow children to have multiple parents, and even allow a child to be its own grandparent. Thus, we can consider the hierarchical to be a subset of the network model. We remove much of the structure of the hierarchy altogether and we let our most common, or the highest resolution entity, become the basis for our model. While this is true geographically, in a non-map context it does not matter if we do not know the exact extent of a particular area, as long as we store pointers from that entity to everything that incorporates it. This model has won a fair degree of acceptance as a data structure for geographic information, although the query ability of these systems may be as limited as that of hierarchical systems (Aronson, 1987).

7.2.3 The Relational Data Model

Returning to the context mentioned above, of levels of development of data management within cartographic and geographic analytical systems, we have now reached the fourth-generation system. In some fourth-generation systems, the data model is termed the relational model. The basic unit within a relational system is the table. Within each table we store the name of an entity, the type of that entity, and the attributes associated with that entity. Each table closely resembles the flat file mentioned above.

One of the attributes associated with a cartographic object, a line for example, would be the points it contains. Within a relational structure, the points would be stored sequentially in a separate file, since normally relational systems do not use ordering of records. The link between the files is by a *key* or index. The index could be a line identifier. So, for example, in a DLG file we could have a particular line representing a part of an interstate highway between two junctions. On the map, the line would be symbolized as a red double line with a black border, a fact which could be stored along with the line attributes. The attributes of the line, however, need not be stored with the line. The line could be numbered with a key, and the corresponding key link to another file could contain the attributes of the line, for example the fact that it is an interstate highway, the type of road surface, the volume of traffic, etc. An additional attribute could also be a further index, for example links (keys) to a file of areas such as counties through which the road passes. At this level, links can be established freely between any of the objects, and only the minimum number of files is required to store the data.

A graphic data structure with relational characteristics is very versatile. We can come into this file structure anywhere, at any point, and retrieve useful information. For the interstate highway, we may have a single arc record and wish to make a query at the arc level. One thing we might be interested in is what arcs meet this particular arc at the end of this line segment. For example, a route-finding application may need to provide in-vehicle instructions based on the connecting highways. All we need do is follow the segment along to the end, sort the arc attribute file, and from that retrieve the polygon identifiers which contain these arcs, and so find the polygons which include a particular location. Thus the in-vehicle system could notify the driver as he passes from town to town, city to city, or county to county.

The relational structure involves some duplication of information between data files. Also, we end up doing a tremendous amount of indexing. The common element within relational structures is the key, sometimes called the index, which is a unique identifier shared between two files. In relational data-base management systems complete software systems take care of retrieval of records based on their attributes, and even contain systems for setting up new tables and entering data into the tables, called data definition languages. Many such systems are now commercially available, and several have been used effectively for geographic data. Relational data-base management systems also minimize data volume by a process called weeding. As we build new files, and it is very easy to build files out of existing files in relational systems, the system automatically looks for duplication of keys within a file and eliminates redundant records.

Query within a relational system allows some very powerful and flexible searches, for example the joining of data sets to provide temporary sets useful for a particular query. Aronson (1987) noted that the relational structure is most amenable to geographic data because of its simplicity, flexibility, efficiency of storage, and its non-procedural nature, i.e., its free-form query. The relational data model has emerged as "the dominant commercial data management tool of the eighties" (Aronson, 1987).

7.3 A REVIEW OF RECENT NON-MAP STRUCTURES

7.3.1 The Hypergraph Structure

The way the relational data model has found its way into GISs is generally in its extended form as the *entity-relationship model*, which is now the basis of several successful GIS systems, for example, the ARC/INFO system (Morehouse, 1985). To some degree, the entity-relationship concept is similar to an idea by Francois Bouille, a French computer scientist who designed the hypergraph approach and successfully introduced the idea into the English language literature on GIS. The approach suggested that a body of theory from computer science and relational data-base management theory could be linked by two mathematical subfields, set theory and topology, then applied to geographical data.

Bouille proposed in 1978 the four fundamental geographic concepts of objects, class, attributes, and relations (Bouille, 1978). We can think of these as cartographic objects, dimension, attributes, and linkages. In addition, Bouille used six abstract data types: object class, attribute of object, attribute of class, relationships between objects, and relationships between classes. Under this system, it was possible to develop an n-dimensional diagram which explicitly stated all of the possible links between the objects. Its expression in topological space was a concept borrowed from the mathematical field of topology, called a hypergraph.

A hypergraph is a network consisting of nodes and edges. The hypergraph provided the mathematical basis for making relations link objects, classes, and their attributes. The entity-relationship model uses similar ideas, as will be seen below. There are examples of implementations of the hypergraph structure, to road networks, for example (Rugg, 1983). The idea led to considerable activity in publishing and research in the area of relational data-base modeling for spatial data.

7.3.2 The Entity-Relationship Structure

Nyerges introduced into GIS and cartography a model developed by Chen (1976) which Chen termed the entity-relationship model (Nyerges, 1980). A very similar model was proposed in the same year by the computer scientists Shapiro and Haralick (1980), and placed into actual use by 1982. This represents a very short gap between research and implementation of new systems, a sign of a vigorous research frontier.

Chen noted that his model was more powerful than the simple relational model, and had all the capabilities of both the network and the relational models. The entity-relationship model has become the standard for many recently developed GISs. In many respects, this is the "single super-flexible structure forming the basis of geographic data management, display and analysis" (Clarke, 1986). An extension of the entity-relationship design is the relational model with non-fixed-length sets of attributes. In computer science, such data management has become called object-oriented, and a number of systems use the approach. As yet, only experimental versions of GISs use this method, although the concept is pertinent to many previous designs.

7.3.3 Hybrid Structures

Roger Tomlinson in 1978 argued that geographic data will never conform to the computer science data models, and that we should not try to force geographic data into structures that were designed for handling non-map, in other words, simple attribute, data. Tominson noted that in mapping and analysis we have two different goals: in mapping, all we really need to do is retrieve the entities and the geographic locational attributes associated with those attributes. In analysis we very often want to retrieve both the entities and the relationships between the entities. Tomlinson said that it might be more appropriate to develop entirely new data models from a geographic viewpoint. A "geographic" data model has yet to be developed. As far as cartography is concerned, such a model may never need to exist. Only in GISs do we need to embed more information than we need to graphically depict a map as a set of symbols, and support cartographic analysis. A "geographic data model" may not be the answer as far as GIS is concerned because no matter how complex the data structure, there are always queries that reveal weakness in the data structure for a particular application.

An alternative is to look at some other approaches that have less flexibility. One approach has been to propose systems with "artificial intelligence," or at least with "expert knowledge," which interact in a more effective manner with the GIS user and structure the data accordingly (for a survey, see Robinson and Frank, 1987).

Finally, we could use the existing approach, which seems to be to build a new or modified data structure for each new application. While the off-the-shelf nature of data-base management systems has made this increasingly easy over the last few years, it is the nature of the map to non-map data link which is of most importance. Systems which support real geographic query allow the user to interact with the data through the map, rather than though a menu or command-line type of interface. A cartographic "window" into the data is essential, as is the ability to view and manipulate cartographic symbols while managing the data.

Clearly, non-map data structures are essential to GISs. On a continuum, they are also vital to analytical cartography, where the management of attributes is critical, but less so for computer cartography, where it is sufficient simply to retrieve the correct attribute. It is appropriate, therefore, to conclude this section on the representation of cartographic data by examining the utility of the non-map data models for analytical and computer cartography.

7.4 DATA STRUCTURES FOR ANALYTICAL AND COMPUTER CARTOGRAPHY

Much of the discussion above has been appropriate for GISs. What about analytical and computer cartography? Automated cartographic systems usually need only to retrieve entities and display them. The most important attributes are locational, and the need for complex statistical operations is minimal. For many map applications, we would be quite happy with a very simple non-map data structure as far as symbolization is concerned. Since more sophisticated cartographic applications such as dynamic maps and vehicle navigation systems require some kind of non-map structure, the two have to be irreversibly linked. Much of the structure of a map is almost totally independent of the locational attributes that a map portrays.

Analytical cartography, and also computer cartography, should concern itself with the management of non-map as well as map data. While the map is the primary vehicle for the communication of geographic properties and distributions, it is the effectiveness of the analytical operation, the combination of data structure and cartographic transformation, which increasingly influences which map the map reader has access to. A system which produces the wrong map, or a sub-optimal map, for the task at hand may be as poor a piece of cartographic software as that which uses poor design or inappropriate symbols.

The power of both map and non-map data structures, as far as analytical and computer cartography is concerned, is in how readily they support cartographic transformations, a theme discussed in the next section. For many reasons, particular cartographic transformations, such as particular symbolization methods, require data to be in a specific data structure. In this case, we can add to the power of our structure by choosing those which allow us to transform between structures with minimal error and data loss.

Chapter 8 introduces the cartographic transformation, and Chapter 9 extends the idea to cover the four major mapping transformations. A particularly important chapter, given the discussion above, is Chapter 10, in which transformations between structures are considered in more detail.

7.5 REFERENCES

ARONSON, P. (1987) "Attribute handling for geographic information systems," *Proceedings, AUTOCARTO 8*, Eighth International Symposium on Computer-Assisted Cartography, Baltimore, MD, March 29-April 3, pp. 346-355.

BOUILLE, F. (1978) "Structuring cartographic data and spatial processes with the hypergraph-based data structure," in G. Dutton (Ed.), *First International Symposium on Topological Data Structures for GIS*, Laboratory for Computer Graphics and Spatial Analysis, Harvard University, Cambridge, MA.

CHEN, P. P.-S. (1976) "The entity-relationship model—toward a unified view of data," *ACM Transactions on Database Systems*, vol. 1, no. 1, pp. 9-36.

CLARKE, K. C. (1986) "Recent trends in geographic information system research," *Geo-Processing*, vol. 3, pp. 1-15.

MOREHOUSE, S. (1985) "ARC/INFO: a geo-relational model for spatial information," *Proceedings, AUTOCARTO 7*, Seventh International Symposium on Computer-Assisted Cartography, Washington, DC, March 11–14, pp. 388-397.

NYERGES, T. L. (1980) "Representing spatial properties in cartographic data-bases," *Proceedings, ACSM Technical Meeting*, St. Louis, MO, pp. 29-41.

ROBINSON, V. B., AND FRANK, A. (1987) "Expert systems applied to problems in geographic information systems: introduction, review, and prospects," *Proceedings, AUTOCARTO 8*, Eighth International Symposium on Computer-Assisted Cartography, Baltimore, MD, March 29-April 3, pp. 510-519.

RUGG, R. D. (1983) "Building a hypergraph-based data structure: The example of census geography and the road system," *Proceedings, AUTOCARTO 6*, Sixth International Symposium on Computer-Assisted Cartography, Ottawa, Ontario, October 16-21, vol. 2, pp. 211-220.

SHAPIRO L. G. AND HARALICK, R. M. (1980) "A spatial data structure," *Geo-Processing*, vol. 1, no. 3, pp. 313-338.

TOMLINSON, R. F. (1978) "Difficulties inherent in organizing earth data in a storage form suitable for query," *Proceedings, AUTOCARTO 3*, Third International Symposium on Computer-Assisted Cartography, San Francisco, CA, January 16-20, pp. 181-201.

VANROESSEL, J. W. AND FOSNIGHT, E. A. (1985) "A relational approach to vector data structure conversion," *Proceedings, AUTOCARTO 7*, Washington, D.C., March 11-14, pp. 541-551.

WAUGH, T. C. AND HEALEY, R. G. (1986) "The GEOVIEW design: a relational database approach to geographic data handling," *Proceedings, Second International Symposium on Spatial Data Handling*, International Geographical Union Commission on Geographic Data Sensing and Processing and the Intenational Cartographic Assosciation, Seattle, WA, July 5-10, pp. 193-212.

8

A Transformational View of Cartography

8.1 A FRAMEWORK FOR TRANSFORMATIONS

Much of the focus of the previous chapters of this book has been upon the tools of mapmaking or upon the structure of cartographic data at each stage in the mapping process. This approach has concentrated upon the *states* of the cartographic data as the data moved from cartographic entity to cartographic object to cartographic symbol. But this is only half of the story. It is time to ask how we move between states, and therefore it is time to concentrate upon the cartographic transformation, and to take *a transformational view of cartography*. In this section we will concentrate upon the *analytical* aspects rather than the computer. Cartographic transformations are central to what analytical cartography is all about. In fact, analytical cartography could be called a discipline in transformations.

Cartographic transformations come in many forms. There are transformations of attribute data, transformations of the locational properties of maps, graphic transformations, transformations of the information content of maps, and the scale transformations of generalization and selection. As will be shown, it is particularly interesting to ask if any given transformation is invertible, i.e., whether or not a cartographic transformation can be reversed to produce the initial starting conditions. Recent advances in mathematical theory have focused attention on transformation inversions which are unstable, i.e., the inverse transformation produces chaos. A thorough knowledge of stable versus unstable cartographic transformations would go a long way toward providing a theory of cartographic transformations. Stable transformations are controllable, and therefore are effectively programmed and modeled, especially with respect to the error introduced. The analytical cartography introduced in this section is the core of

132

what stands behind much of the computer cartographic software developed to date. This means that it will also be reflected in the computer cartography of the future.

Many of the ideas in this chapter have been adapted from the paper "A Transformational View of Cartography", in which the cartographer Waldo Tobler stated his conceptual viewpoint on analytical cartography (Tobler, 1979). Tobler's idea of transformations originated with a single set, the transformations of map space caused by map projections, that is, point-to-point transformations. This was expanded to cover other dimensions, and transformations between dimensions. To these can be added transformations of scale, and the mapping transformation which actually produces the map, which is known as the symbolization transformation. In this and the next chapter, these four types of transformations will be examined in detail and used to present a transformational view as a unifying theme in analytical cartography.

The idea of transformations in cartography has roots in the work of the cartographer Arthur Robinson, who devised a way of classifying the types of cartographic symbolization and map types in common use (Robinson and Sale, 1969). Robinson placed methods into the framework of a set of states classified by level of measurement and dimension (Figure 8.01). Robinson proceeded to place most of the methods used to make maps into one or other of the categories. David Unwin expanded upon Robinson's classification to add the distinction between map types and data types. Map data transformations and map-type transformations are then distinct. Data attributes define data types, while the type of symbolization defines the map-type. This implies that cartography involves a data type-to-map type *symbolization* transformation. Analytical cartography is the scientific treatment of the properties of the data to map-type transformation.

Content Scaling Level	Defining Relations	FORM OF CARTOGRAPHIC SYMBOL		
		POINT	LINE	AREA
Nominal	Equivalence	Wholesale and Retail Establishments	Highway Connectivity	Land Ownership
Ordinal	Equivalence Greater than	Small Medium Large Population Centers	Roads by Degree of Improvement	Yield
Interval	Equivalence Greater Than Ratio of Intervals	45 89 72 60 42 Spot Elevations	Latitude/ Longitude Grid	Date of First Settlement
Ratio	Equivalence Greater Than Ratio of Intervals Ratio of Scale Values	Area Proportional to Population	Population Density Isopleths	Darkness Proportional to Population Density

Figure 8.01   Classification by Scaling and Dimension (after Robinson and Sale, *Elements of Cartography, 3e,* copyright © 1969, by John Wiley & Sons, Inc. Used with permission.)

To start with Robinson's classification, the primary concern was with the symbolization transformation, and the idea was to break down the ways in which symbolization takes place by two factors, the type of data and the level of attribute measurement. In this context, the symbolization method means which actual symbols are used on a given map; blue tints, brown lines, dot patterns, etc. The map type is the cartographic technique used to produce the map. As an example, a map with areas shaded red or blue depicting the winning political party within counties has a data type which is area/nominal, and a map type which is colored areal. The fact that we used red and blue, and shaded the areas within the boundaries of the counties, is the result of the symbolization technique. Under the Robinson/Unwin system, the *states* in the mapping process could be summarized as area/nominal data to colored area map to, let's say, color ink-jet print from a particular choropleth mapping package. The sequence is *data type* to *map type* to *symbolization* to *map*.

The framework for cartographic transformations proposed here is that the mapping process is really the sum of a series of state transformations and their interactions. At any stage in the process, the cartographic information with which we are working has a data structure, and the transformations operate upon this structure. By using Robinson's types of symbolization and Unwin's data and map types, we can lay out a finite set of transformations, each of which can then be examined in more detail. Before we go into the *how*, however, we should be fully aware *why* we are performing these important transformations.

8.2 REASONS FOR TRANSFORMING CARTOGRAPHIC DATA

Three of the most important reasons for transforming cartographic data are generalization, converting the geometry of the map base, and changing the data structure. The first reason is *cartographic generalization*. Generalization underlies much of analytical cartography. A good subtitle for the discipline of cartography would be a "discipline in reduction", since we are always changing map scales from larger to smaller as we move from data to map and attempting to understand what happens in the process. We may wish to have maps collected in different ways transformed to a common basis. It may be a common statistical basis, where we have directly measurable quantities of the same values, where the same statistic, for example, needs to be compared across different countries. Or we may want maps put onto the same geographic reference base; for example, we may want maps with different map projections transformed to one common coordinate system. Another reason for transformations is that cartographic analysis, modeling, and different map types and symbolization methods all favor a particular cartographic data type, so if we want to do automated contouring, we usually somehow have to convert our data into integers in a grid data structure, because this is how about half of the automated contouring packages work. The problems of transforming cartographic data are usually as important as the problems associated with making the final map.

Another major reason for cartographic transformations is *to change data structure*. The different phases that we go through when we handle cartographic data, such as data input or output, and data manipulation of various different types, all have their best structures. The best way to digitize a polygon map for choropleth mapping, unless you want to duplicate information, and assuming that one wants the capability to do some editing of the data after initial entry, is to enter the data as connected line segments forming chains, and then to deal with the problem of transforming them into the format which we really want, which in this case is the polygon list, so that we can shade the areas, as required by our symbolization method. Transformation between data structures is worthwhile because the real difficulty is digitizing and we want to make this stage as easy as possible. Data structure transformations, however, are for the purpose of suiting the cartographic data for different types of analyses or symbolization, and are not an end in themselves.

Many other processes involve exactly the same issue; for example, there are certain analytical operations which are far better performed on a grid rather than on a vector data structure, and vice versa. We almost always find ourselves needing to transform between data structures, so that the data structure transformation is usually the first, if not the most important transformation we may apply during the mapping process.

8.3 TYPES OF DATA

The basis of Robinson's classification of symbolization and map types was the division by data dimensions: the familiar division of cartographic objects into the types *point, line,* and *area*; to which can be added *volume*. Point data can be described simply by a location with an attribute. Line data are usually thought of as being a string of connected straight-line segments, but can also be represented by a mathematical function such as a polynomial or a spline. Area data are regions or polygons, predefined geographic areas with a boundary, a given topology, and an attribute, such as the state of New York. Volumes can represent either continuous statistical distributions, such as air pollution, or terrain elevations.

The second division of cartographic objects under Robinson's classification is by the level of measurement of the data. *Level of measurement* means the classification of the non-map data by its complexity. This division was defined by Stevens (1946). The normal categories we use are *nominal, ordinal, interval,* and *ratio. Nominal* simply means assigning a name or label. A nominal cartographic feature may be a place name associated with a point object, such as the label *New York* assigned to the location 40 degrees 40 minutes north latitude, 73 degrees 58 minutes west longitude on a small-scale map. Such objects may have rules or methods associated with them, especially in relation to how they are cartographically symbolized and what map types can be used with them.

Ordinal data has a sequence or ranking associated with it, so relations like "larger", "smaller", or "greater than" can be used. *Interval* data has a measured numerical value as an associated attribute, such as a measured field elevation. The scale, however, is not absolute but is pegged to an arbitrary unit or false zero. Thus a geological map may classify rock types by age before present (BP), representing an age which cannot be precisely determined, yet gives a relative history to the rock sequences which is less arbitrary than a set of younger than/older than references. A *ratio* value is a measured value on a scale with a meaningful zero on which mathematical operations can be performed. The Kelvin scale, for example, has a zero determined by the lowest possible temperature rather than the temperature at which ice melts. Examples of ratio values are population density, change rates, and percentages. Transformations between levels of measurement are not necessarily map data transformations, but they do affect the types of maps we can use to portray the transformed data. A good example, within a normal cartographic representational procedure, is in making a choropleth map. We may collect data on population (ratio) and the areas of states (ratio). From these two ratio values we compute a third, in this case population density in persons per square kilometer. This is a simple numerical, mathematical transformation, and is invertible on the map since we will be showing the outline of the states and therefore allowing the computation of their areas (if we show a scale). Next we represent our ratio value using an area technique known as choropleth mapping, which first reduces the ratio data to ordinal by classification, and then assigns ordinal shade categories and patterns to the regions on the map. The entire set of level-of-measurement transformations is depicted in Figure 8.02. Normally, the symbolization transformation is not an invertible transformation, and frequently is far from simply performed, since there are many possible ways to subdivide the data into categories.

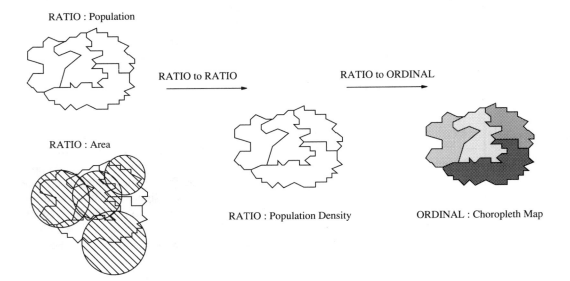

Figure 8.02   Level of Measurement Transformations for Choropleth Mapping

Jenks and Caspall (1971) provided a means of measuring the error involved in this last data transformation, allowing an objective evaluation of the various classification methods. In fact, Jenks and Caspall presented four measures of the accuracy of this transformation. One of these was a composite of the remaining three, which were all based on comparison with an idealized model of a choropleth map with enough classes to have one class per data value. The error was computed using the concept of a three-dimensional solid raised to the height of the choroplethic data value and the statistical standard deviation. With only one choropleth value (the worst case) the entire map would be represented by the data mean. The overview error would then be the sum of the absolute values of the data values minus the mean, in each case times the areas represented on the map. Similarly, the tabular error is simply the height differences, without the transformation to volumes. The two versions of this measure are the absolute and relative values of this number. The error indices provide an objective function to maximize, and Jenks and Caspall presented a worked example of a classification optimization for the choropleth data they used. An optimal set of choropleth mapping transformations would minimize this error.

Thus for a choropleth map we can transform the data into a map, and simultaneously make definitive statements about the error or information loss in the transformation, therefore defining the invertibility of the transformation. These types of transformations, their characteristics, how they can be performed, their invertibility, and the error both spatial and aspatial which they embody form the bulk of the subject area for analytical cartography. So familiar are these transformations that we often perform them without thinking, or even perform them mentally while looking at a table of numbers or a map.

8.4 TYPES OF MAPS

David Unwin presented a division of the Robinson classification into data types and map types. Where Robinson was interested in cartographic symbolization, Unwin was interested in spatial analytic aspects of the state descriptions. Unwin presented two three-by-four tables, shown as Figure 8.03. The classification was the same as Robinson's, with the exception that volumes (surfaces) were added as a data type. The division into map type and data type was based on the distinctions between cartographic entities and their symbolization methods. Thus under Unwin's schema, a nominal linear data type was a road, and a nominal linear map type was a network map. The road could be described as the cartographic entity. The network map, however, is both a type of symbolization and a type of cartographic object. In the following discussion, we will distinguish between the transformation of a cartographic entity to a cartographic object via geocoding and the adoption of a cartographic data structure, and the eventual symbolization transformation of the object required to make a map. The emphasis is therefore not on map types or data types, but on the transformations between the states into which cartographic entities, objects, and symbols fall.

DATA TYPES		Point	Line	Area	Volume
Nominal		City	Road	Name of Unit	Precipitation or soil type
Ordinal		Large City	Major Road	Rich County	Heavy precip. Good soil
Interval		Total Population	Traffic Flow	Per Capita Income	Precip. in mm or Cation Exchange
Ratio					

MAP TYPES		Point	Line	Area	Volume
Nominal		Dot Map	Network Map	Colored Area Map	Freely Colored Map
Ordinal		Symbol Map	Ordered Network Map	Ordered Colored Map	Ordered Chromatic Map
Interval		Graduated Symbol Map	Flow Map	Choropleth Map	Contour Map
Ratio					

Figure 8.03   Map Data and Map Types (after Unwin, 1981)

This approach, the transformational view, is summarized in Figure 8.04. The domain of interest of the analytical cartographer is the full set of state transformations possible, yet it is also obvious that there will be an optimal set or pathway of transformations to make a particular map. The symbolization transformation, in which the map is realized and finds a physical description and form, forms the topic of computer cartography, being by definition technology dependent. Under the transformational approach, four major cartographic transformations seem to shape the way in which mapping takes place. These are: the geocoding transformation between cartographic entities and cartographic objects, involving transitions between levels of measurement, changes in dimension, and changes in data structure; changes in map scale; changes in the locational attributes of the data, i.e., transformations of the map base itself; and the transformation resulting in symbolization and a map.

Finally, we should note that transformations of cartographic objects can yield objects with fewer dimensions, or even no dimension at all (scalars). The intersection of two lines produces a point, and the measurement of area of a polygon produces a scalar. Cartographic measurement, part of cartometry, is itself simply a transformation to a lesser dimension object, or a simpler geometry.

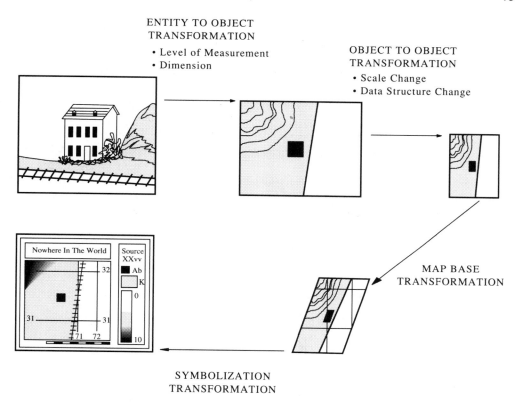

Figure 8.04   The Four Major Mapping Transformations

8.5 TRANSFORMATIONS OF THE MAP SCALE

To cartographers, the range of interest is the map scales between about 1:1,000 up to about 1:400,000,000. Remember that *large scale* means a representative fraction with a smaller number as the denominator, and covers a small area in a great deal of detail. A 1:1,000 map is approaching a plan, and is of as much interest to the surveyor or the engineer as to the cartographer. Similarly, a small-scale map has a large number in the denominator of the representative fraction, and shows a large area in less detail, perhaps a whole state, country, or even a continent.

To provide some examples of these scales, at the large-scale end 1 meter on the ground is 1 millimeter on the map; while at the small scale end the entire earth map would fit onto the surface of a golf ball. We have different names for the disciplines concerned with scales beyond these ranges. At extremely small scales, the concerned disciplines are astronomy and planetary physics; while at extremely large scales the fields include planning, surveying, and engineering. The scale of 1:1 is called reality, and scales beyond this are the enlargement disciplines, chemistry, microbiology, and particle physics. We use different instruments at different scales, and the scales affect the collection of data with those instruments. We cannot, for example, measure interplanetary distances with a ruler; neither can we observe distant galaxies with an electron scanning microscope.

Cartographically, transformations between scales are almost always from larger to smaller scale, and proceed by a process of *generalization,* involving the selection of features to survive the scale change, the deletion of detail, the smoothing or rounding of the data, and sometimes the change in symbolization necessary to deal with the new scale, which can even involve moving objects to facilitate interpretation. Some automated methods for scale changing are discussed in Leberl et al. (1986). Figure 8.05 summarizes the types of scale change discussed by Weibel (1987) in the specific context of digital elevation models but applicable to most types of cartographic data. Weibel noted that cartographers perform four basic operations while generalizing cartographic data. These are elimination (selection), simplification, combination, and occasionally displacement.

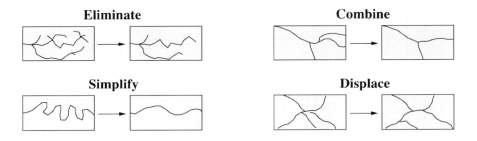

Figure 8.05   Scale Change Transformations

Elimination involves the dropping of cartographic features as we move from large- to small-scale maps. Simplification involves classical generalization, such as placing a dot for a city on a small-scale map or showing a river as a smoother and smoother line at smaller and smaller scales. Combination involves joining features together at smaller scales, for example joining a river with its tributaries, or combining island groups into a smaller number of "representative islands". Displacement is occasionally necessary to reveal structure or to allow space for labels. This is often the case when many features are clustered together in a single region on the map.

Our scientific ideal is always to collect data at scales which are larger than we need for a particular map so that we can eliminate, simplify, and generalize using known rules, methods, and criteria. Much of manual cartography, and a great deal of computer cartography, has to do with sets of rules for determining the generalization transformation. Probably the best defined are the rules for generalizing a coastline. We generalize through scales, gently smoothing the coastlines, but we also make jumps at certain scales. When an island becomes a dot at a particular scale we may choose to eliminate it. Similarly, whole inlets or river estuaries may disappear on small-scale maps. Figure 8.06 shows this phenomenon. The lake in question actually appears to change shape as it moves through the scales at which it is mapped.

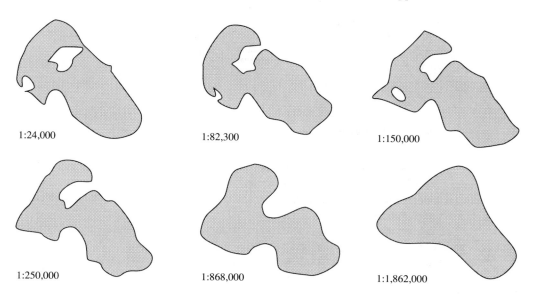

1:24,000   1:82,300   1:150,000

1:250,000   1:868,000   1:1,862,000

Figure 8.06   Copake Lake, New York, at Various Map Scales.

Only recently have cartographers thought about the inverse transformation between scales. This is probably because we used to call this process "cartographic license" or sometimes "witchcraft". A better name would be *enhancement*, and it is important to note that not all enhancement is witchcraft, for we may have good information at some places on the map about patterns at larger scales. Extending the characteristics of this variation uniformly over the map artificially may be termed *emphatic enhancement*. On the other hand, using a model to generate variation, such as fractal enhancement of coastlines, is *synthetic enhancement*. Dutton (1981) was a pioneer in this work and his ideas have since been extended to other types of data (Figure 8.07).

8.6 TRANSFORMATIONS AND ALGORITHMS

The most elegant statement of the characteristics of a transformation is in a mathematical equation. Invertibility of equations can be proven by algebra or other methods, so that we can rearrange them to find other variables given a different set of starting conditions. When dealing with cartographic transformations, we do indeed have many transformations which are expressible in the language of mathematics. Others, however, are less amenable to such simple description. In computer science, an *algorithm* is

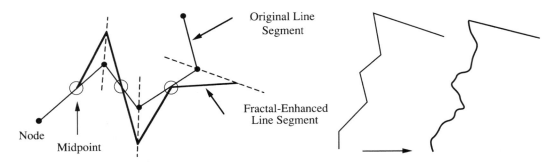

Figure 8.07   Dutton's Fractal Enhancement for Coastlines

a special method of solving a problem. An algorithm may be stated either as a simple formula, or as a set of sequential instructions to be followed to arrive at a solution. Church's theorem implies that if a process can be so stated, then it can be automated, in this case programmed.

The set of algorithms for cartographic transformations operate on cartographic data which has some preexisting data structure, and result in a form of the data which is closer to the map required. A good summary, with apologies to Wirth, would be that

cartographic data structures + cartographic transformational algorithms = maps

When you dig down far enough within any piece of cartographic software, however complex, cartographic transformational algorithms are the very nuts and bolts with which that piece of software is constructed. Similarly, the cartographic data structures involved are the materials which the transformations hold together to produce maps. An understanding of the data, the problems, the data structures, and the algorithms is really all the analytical cartographer needs, except of course a real desire to make *better* maps. In the following chapters of this section on analytical cartography, we will go closer into the algorithms which perform cartographic transformations, and show how some simple computer programs can be written to implement the transformational approach taken here.

8.7 REFERENCES

DUTTON, G. H. (1981) "Fractal enhancement of cartographic line detail," *The American Carto-grapher*, vol. 30, no. 1, pp. 23-40.

JENKS, G. F. AND CASPALL, F. C., (1971) "Error on choroplethic maps: definition, measurement, reduction," *Annals, Association of American Geographers*, vol. 61, no. 2, pp. 217-244.

LEBERL, F. W., OLSON, D. AND LICHTNER, W. (1986) "ASTRA—a system for automated scale transition," *Photogrammetric Engineering and Remote Sensing*, vol. 52, no. 2, pp. 251-258.

ROBINSON, A. H. AND SALE, R. D. (1969) *Elements of Cartography*, Wiley, New York, Third Edition.

STEVENS, S. S. (1946) "On the theory of scales of measurement," *Science*, vol. 103, pp. 677-680.

TOBLER, W. R. (1979) "A transformational view of cartography," *The American Cartographer*, vol. 6, no. 2, pp. 101-106.

UNWIN, D. (1981) *Introductory Spatial Analysis*, Methuen, London.

WEIBEL, R. (1987) "An adaptive methodology for automated relief generalization," *Proceedings, AUTOCARTO 8*, Eighth International Symposium on Computer-Assisted Cartography, Baltimore, MD, March 29-April 3, pp. 42-49.

9

Map-Based Transformations

9.1 TRANSFORMATIONS OF OBJECT DIMENSION

In Chapter 8, we met the transformational view of cartography. Two types of transformations were considered: first, transformations between the types of data and types of map involved in the mapping process; and second, transformations between map scales. A third type of transformation is the transformation of cartographic objects themselves via their dimensions. Since these are transformations of either the locational aspects of the data or of the structure used to represent a cartographic object spatially, they deal centrally with the core of map information and are close to the heart of analytical cartography.

We have already noted the contribution of Tobler's transformational view (Tobler, 1979). Tobler saw cartographic object dimensions as states for cartographic data, and enlarged the set of state transformations beyond the classical point-based cartographic conversions, to include the other dimensions with which we specify locations. He envisaged a three-by-three table—of points, lines, and areas versus points, lines, and areas—which contained nine possible groups of transforms, with many specific cases and divisions in each category. With volume added as a cartographic object, we have 16 possible sets of transformations, arranged in a four-by-four table. Tobler saw this table as a transformation matrix, a set of possible transformations between states at different times. Within this viewpoint, time 0 and time 1 are states, and the cartographic data goes through a transformation between states (Figure 9.01).

Changes within this transformation matrix along the diagonal are transformations either between scales or between data structures. The final transformation is that between the cartographic data and a map, the symbolization transformation. This chapter will focus upon the dimensional transformation, and finish by discussing the

144

State at Time One

	Point	Line	Area	Volume
Point	Point to Point	Point to Line	Point to Area	Point to Volume
Line	Line to Point	Line to Line	Line to Area	Line to Volume
Area	Area to Point	Area to Line	Area to Area	Area to Volume
Volume	Volume to Point	Volume to Line	Volume to Area	Volume to Volume

State at Time Zero (vertical label at left)

Figure 9.01   The 16 Single-Step Dimensional Transformations

symbolization transformation. Chapter 10 follows up specifically with a more in-depth discussion of scale transformations and data structure conversions.

What we would like to be able to do as analytical cartographers is to express a cartographic transformation either as an equation, or as an algorithm that allows us to project state 0 onto state 1 in a fully described way. The transformation from state 0 to state 1 has an inverse transformation which transforms from state 1 to state 2, equivalent to state 0. If such a transformation exists, the entire transformation belongs to a special subset of transformations known as invertible transformations.

In this chapter, we shall deal with transformations at the low-dimensional end of the transformation matrix, especially point-to-point transformations. Most of these are transformations within a single type but between scales, and of non-map data scaling. Some of them have as their input a cartographic line, and as their output a measurement based upon properties of the line. Not all of these are invertible transformations. Point-to-point transformations, which Tobler called the classical cartographic transformations, are very central to analytical cartography. An understanding of point-to-point transformations is essential to analytical cartography, more essential than many of the other dimensional transformations. Map-based transformations are pivotal to understanding the ability of analytical cartography to make systematic inquiries into how data structures representing cartographic objects share space. Analytical cartography seeks to determine how the geographic properties of the space sharing can be used in analysis, modeling, and prediction.

The transformation matrix shown in Figure 9.01 shows only the single-step transformations. It is possible also to have multiple-step transformations, and these are indeed common. As an example, if we measure elevations in the field, it is most convenient to record them at sets of irregularly distributed points. We can describe these points digitally as point locations with the attribute of elevation. The data value we have sampled, topography, is volumetric, of which we have sampled only the surface with sets of point data observations. We take a continuous value, distributed over a volume, we sample using a series of points with attribute values associated with them, and we then transform the attributes of the points into intersection points on a regular grid based on the spatial distribution of points (Figure 9.02).

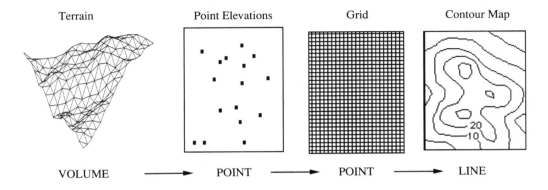

Figure 9.02   An Example of a Multi-Step Transformation

Our motive may be, perhaps, that we want to use an automated contouring technique requiring data in the grid structure. We transform from time 0 to time 1, and then on to time 2. Time 1 to time 2 is a point-to-point transformation, where the first set of points is irregularly spaced, and the second set of points falls at the vertices of a regular grid. We might take the series of points and use it to feed a set of lines through the points to make contour lines. So we have input a series of points, transformed them to another set of points, and then transformed them to a set of lines. We then draw the lines on the map as contours and use them in the map reader transformation to communicate the impression of topography, a volume.

Using this transformational view of cartography, we have a versatile framework for analytical cartography, and we can place any kind of cartographic data manipulation into this framework. In the remainder of this chapter, we will take one set of transformations and start working through specific examples. These are transformations of the *locational attributes*, in other words, coordinate transformations. This type of transformation changes the geographic space of the base map itself. For small-scale maps, with such transformations we are largely talking about *map projection* transformations.

9.2 MAP PROJECTION TRANSFORMATIONS

We can have three approaches to transforming the coordinates of small-scale maps. Approach number 1 is to ignore the problem of the earth's shape, and unfortunately this is a frequently employed solution to the problem. Ignoring the projection results in some unexpected and inaccurate spatial statistics and poor maps, especially if we take maps which have been prepared on different map projections and overlay them. Many mapping agencies simply map 1-degree by 1-degree cells onto a single display square. At middle latitudes, this seems to work fine, but as we approach the poles, the north-south distortion becomes sufficient to make computations of areas, directions, and bearings meaningless.

The second approach to coordinate transformations is to use statistical techniques such as rubber sheeting or adjustment. Surveyors have applied adjustments to measured coordinates for many years, distributing the random error of measurement proportionately around points. In remote sensing, images are transformed into map space using the same principle, that of a least squares fit of the image to the map space. Under ideal circumstances we would know every geometric property of the map, and understand the relationship of these geometric properties to latitude and longitude. This would allow us to know all of the equations to perform the space transformation. This is rarely the case. Rubber sheeting is a statistical and empirical approach. It works, it is frequently used, and it is acceptable cartographically as long as we realize that we have incorporated systematic error into our final map and that we have lost some of the cartographic fidelity. Of course, since this method is empirical the transformation is not directly invertible, although it often comes close to it if we retain the control point locations. The transformation itself, similar in many respects to the cartogram, is covered in Section 9.5.

The third approach is to compute the characteristics of the transformation itself geometrically. There are two ways of doing this; we can be precise, or we can be approximate. Which we choose depends largely upon what we want to do with the data. For example, we can choose any one of three geometric forms to model the earth's surface: a sphere, an oblate ellipsoid, or a geoid. The sphere is most familiar to us as the common globe, and is perhaps the easiest to understand. The oblate ellipsoid is a more precise model, allowing higher levels of accuracy. An oblate ellipsoid is the volume traced out by an ellipse rotated about its minor axis. Since the earth is about 42 kilometers fatter across the equator than it is pole to pole, the difference from a sphere is very small. A common estimate of the degree of flattening is 1:294.98, an estimate established by measurement in 1866 by Alexander Ross Clarke, and is still normally used for the mapping of the United States, although it will be phased out as the North American Datum 1983 (NAD83) becomes the standard. Since it was established that different figures could be computed for different parts of the globe, a "best-fit" ellipsoid was chosen, and is given as the 1980 Geodetic Reference System value of 1:298.257, the basis of the NAD83 (Snyder, 1983). The ellipsoid is important, since it makes our maps more accurate and capable of higher precision, but the third earth model is even more precise. In this model, we fit an empirical surface to the earth's gravity field. How far this deviates from the ellipsoid, or usually the local best-fit ellipsoid, depends upon where we are, giving us a lumpy ellipsoid-like figure known as a geoid. How precise we need to get depends on our function. A map of United States population clearly does not need to be precise beyond the best-fit sphere, while the guidance system of a spacecraft needs much more accurate and precise maps.

Cartography traditionally has worked with the sphere since the manual construction of map projections is tedious. The sphere is simplistic, but it makes the calculations much easier. So when we talk about map projections, usually we talk about projections from a point source on a sphere to points on a flat plane. Map projections involve two different types of mappings, both of which benefit from use of the computer. Figure 9.03 shows a representation of a sphere and the various types of possible mappings of points on the sphere onto points on a map.

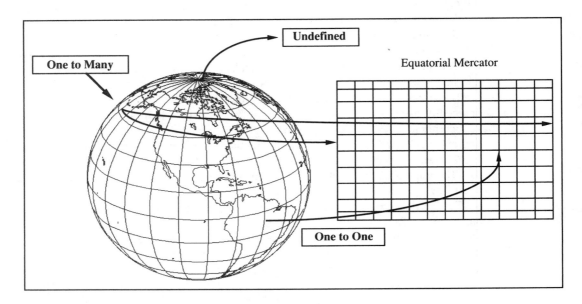

Figure 9.03   Mapping Points on a Sphere to Points on a Plane

A point on the earth's surface maps onto a point on the map. Geometry actually uses the term "mapping" to describe this process, borrowed from cartography. This is usually called a *one-to-one mapping*, and it is preferred because it is well behaved and simply invertible. Imagine an old favorite, Gerhardus Mercator's projection of the earth, one we are all familiar with from the wall of our elementary school classroom. Think about the usual dividing line on the equatorial Mercator projection, 180 degrees east (or west). This single line does not map onto a line; it maps onto two lines, and becomes undefined as we approach the poles. This is an example of a *one-to-many* mapping, and an *undefined* mapping, and these are the kinds of mapping transformations which are extremely difficult to invert.

Different geometries apply in the spherical and planar cases. For example, most people are familiar with the geometry of plane triangles (Figure 9.04). We know things like the law of sines. If a triangle has sides A, B, and C, with opposite angles a, b, and c, then we know that

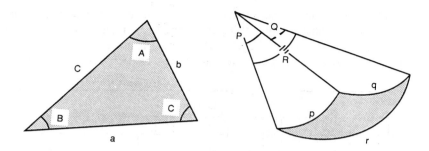

Figure 9.04   Planar and Spherical Triangles

$$\sin(A) \;/\; a = \sin(B) \;/\; b = \sin(C) \;/\; c \qquad\qquad 9.01$$

We often use this and other formulas to solve general purpose triangles. It is a little different on a sphere where we use the spherical triangle. For example, if you are flying an airplane from New York to Chicago, you may want to minimize the amount of fuel you use by flying along the shortest path. To fly along the shortest or great circle route, you would have to fly along the shortest distance on the surface of a sphere between two points. This is a different geometry from the regular triangular case. We have a spherical rule of sines, in which we have

$$\sin(p) \;/\; \sin(P) = \sin(q) \;/\; \sin(Q) = \sin(r) \;/\; \sin(R) \qquad\qquad 9.02$$

since the lengths of our sides are angles, too. If lengths are angles, they are usually expressed in radians rather than in degrees. We usually express latitudes and longitudes in degrees minutes and seconds, but in most spherical geometry we deal with radians, and most of the angles are angles at the center of the earth. In particular, computer programs and calculators often require angles in radians. Under the digital cartographic data standards, the suggested use is the decimal degrees format. Among existing software, the degrees, minutes, seconds format, i.e., DD.MMSS, is frequently encountered. To convert angles in degrees to radians, use the following formula:

$$\text{radians} = 2.0 * \Pi * (\text{degrees}+(\text{minutes}/60)+(\text{seconds}/(60*60))/360.0 \qquad 9.03$$

Let's now set up a spherical angular referencing system, and you will notice that this involves going through a transformation so that you can get a range of values. Figure 9.05 shows a longitude which is an x value going from negative 180 at the international date line to positive 180 at the international date line. Notice that at the date line we have a one-to-many mapping. We have chosen a central meridian, which in this case happens to be the origin of zero at the prime meridian. Going up and down, we choose the equator as zero; we make the north pole 90 and the south pole -90, 90 degrees north and 90 degrees south. Notice that here is another one-to-many mapping. On this map we have one-to-manys around the edge, and one-to-ones within the map itself, where fortunately we can handle most of the transformations without too much trouble. The transformation starts with latitude and longitude pairs, and yields Euclidean coordinates (x,y). With the correct scale factor, the x and y values are actual measurements in map millimeters which we can use in a graphics program when we plot a map. λ and ϕ are longitude and latitude, respectively, given in radians.

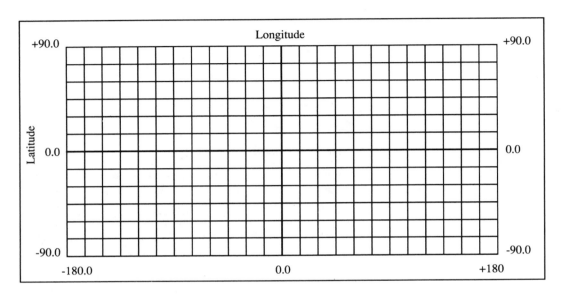

Figure 9.05   Generalized World Coordinate System

Map projection transformations deserve a closer look here before we move on to planar-only transformations. The Mercator projection, which strictly we should call the equatorial Mercator projection, is a cylindrical projection. It is based on a projection of the Earth's surface onto a cylinder wrapped around the earth so that it just touches at the equator and has a central axis which passes through both poles. The formulas for Mercator's projection are

$$x = R \; s \; (\lambda - \lambda_0)$$ 9.04a

$$y = R \; s \; \ln \tan (\Pi/4 + \phi/2)$$ 9.04b

where R is the radius of the sphere on the map, s is the scale factor (representative fraction of the map), and λ_0 is the longitude in radians of the central meridian. Since the radius of the earth at the equator (under the Geodetic Reference System of 1980) is 6,378,137 meters, scale factors have to reduce that to some reasonable value so that the map will fit onto a computer screen or a sheet of plotter paper. At 1:42,000,000 the earth maps onto a globe about as big as a basketball. With **s** set to 0.00000001, the map will have an equatorial dimension of about 0.4 meter. In most Mercator projections the origin is the prime meridian, longitude zero, so most Mercator projections appear with Europe in the middle, or at least with Greenwich in the middle, since this is the origin of the coordinate system. There is nothing which says that we have to focus the projection there, but it is rather convenient, since the international date line becomes the space and time dividing line for the map. Also, the one-to-many mapping at the date line is least inconvenient, since it passes mostly through water. Strictly speaking, the 180th meridian is not the international date line, since the line is moved to avoid land. We can plot world maps using the Mercator projection using these transformation equations. The spacings on the Mercator projection graticule get closer and closer as we approach the equator—the effect of taking the logarithm.

Map projection is a part of analytical cartography where the use of computers is really advantageous since the repetitive computations are very tedious. Most people can work through the formulas here, but it is indeed humbling to realize that Gerhardus Mercator presented the projection in 1569, and that the formulas were published in 1599 by Edward Wright. The following is a C language function designed to perform the equatorial Mercator transformation for a single point.

Function 9.01

```
/* Function to transform a point to Mercator's projection
/* Assumes a point according to the structure POINT
/* presented in Chapter 5
/* LAMBDA_SUB_ZERO is defined in define.h as 0.0
/* PI is defined in define.h as 3.141593
/* R is defined in define.h as 6,378,137 meters
/* S is a scale factor
/* SMALL is defined in define.h as a small real value
/* MAX_NORTH_LAT and MAX_SOUTH_LAT should also be defined and less
     than PI/2
/* #include <math.h> for math functions
/* Function returns value 1 for unsuccessful transformation
/* otherwise 0, with point.x and point.y set to meters
*/
#include "cart_obj.h"
int mercator (point)
struct POINT point;
/* Note that x and y are in radians */
{
    /* Test for undefined latitude to y conversion */
    if ((fabs(point.y) - (PI / 2.0)) < SMALL) return (1);
    if (point.y > NORTH_MAX_LAT) return (1);
    if (point.y < SOUTH_MAX_LAT) return (1);
    point.x = R * S * (point.x - LAMBDA_SUB_ZERO);
    point.y = R * S * log ( tan (( PI / 4.0) + (point.y / 2.0));
    return(0);
}
```

The same transformation can also be expressed using matrix notation. If we start with longitude and latitude, and apply a transformation to them, we get one from the other.

$$\begin{bmatrix} x \\ y \end{bmatrix} = \begin{bmatrix} T \end{bmatrix} \begin{bmatrix} \lambda \\ \phi \end{bmatrix} \qquad\qquad 9.05$$

It also works the other way around.

$$\begin{bmatrix} \lambda \\ \phi \end{bmatrix} = \begin{bmatrix} T \end{bmatrix}^{-1} \begin{bmatrix} x \\ y \end{bmatrix} \qquad\qquad 9.06$$

If the purpose is to get longitude and latitude, we can start with their cartesian coordinates taken by measurement from a map on some map projection and we can apply the inverse of this transformation to yield longitude and latitude. The beauty of this matrix notation is that it often translates directly into an algebra for the solution of the problem which is far more elegant than assignment statements, otherwise known as equations, in a program.

Computer cartographic software to perform the transformations between map projections is increasingly part of the more sophisticated computer mapping packages. Especially important is the ability to transform between coordinate systems based on specific map projections, such as the Universal Transverse Mercator and the State Plane systems. An excellent reference on map projections is Snyder's *Map Projections—A Working Manual* (Snyder, 1987). This work includes references to computer programs to perform the map projections listed and discussed. Also of interest is the NOAA computer package the General Cartographic Transformations Package, a set of FORTRAN programs and subroutines, along with data files containing all the necessary constants for the various standards and datums, as well as constants for the map projections. Twenty map projections, with forward and inverse calculations are included in the public domain package (NOAA, 1988).

9.3 PLANAR MAP TRANSFORMATIONS

In the remainder of this chapter, we will consider some simple geometric transformations, still using map coordinates. Even though some of these transformations are pretty simple, they are also nevertheless powerful. Generally, here we will be dealing with transformations between point cartographic objects and objects with the same or higher dimensions. In most cases, we will discuss a problem, note the mathematical expression required for a solution, and then provide a computer program to yield the answer.

9.3.1 Transformations Based on Points

Distance between Two Points

Let's start with something point-based and simple, the computation of the distance between two points. For example, consider two locations in space; let's call the first (x_1, y_1) and the second (x_2, y_2). Given their locational attributes, otherwise known as their coordinates in space, and assuming planar and not spherical geometry, what is the distance between them? This is not a puzzling problem, having been figured out over 1,000 years ago by Pythagoras. The distance between those two points is equal to the square root of the sum of the squared lengths of the sides of the triangle made between these two points and the orthogonal axes. In other words, the distance is the square root of the difference between the two x values squared, plus the difference between the two y values squared. Pythagoras actually would have said this, that the areas of squares formed on the sides on the triangle made by the two points and a right angle with the axes are such that the largest triangle has an area equal to that of the sum of the smaller two.

Here is a way of getting the distance between two points, which is simply a derived measure:

$$\text{distance} = \sqrt{[(x_2 - x_1)^2 + (y_2 - y_1)^2]} \qquad 9.07$$

but it is easy to apply an objective to it, for example to minimize distance.

Function 9.02

```
/* Function distance : distance between two points */
#include <math.h>
float distance(x1, y1, x2, y2)
     float          x1, y1, x2, y2;
{
     return ((float) pow((x1 - x2) * (x1 - x2) +
         (y1 - y2) * (y1 - y2), 0.5));
}
```

We may have a whole series of points and we want to find which two points are closest together. We have to calculate this value for all pairs of points and then sort the values to find the lowest, which will give us the nearest neighbors. We have already formulated an algorithm for distance computation. Distance minimization is a potential application of this algorithm. We can think of a matrix of points by points, with matrix values as the distances between the points, as being a useful transformation of the map space. In this case, the transformation is invertible, since the surveying practice of trilateration allows us to recompute relative locations from distances.

Length of a Line

Length is very similar to distance. A long line in computer cartography usually consists of a set of joined line segments each consisting in turn of pairs of points with a connecting straight line. The length of each of these segments can be computed using the algorithm above. We simply have to add the lengths of all the segments to compute the length.

$$\text{length} = \sum_{i=1}^{i=npts} [\sqrt{((x_i - x_{i-1})^2 + (y_i - y_{i-1})^2)}]$$

9.08

Function 9.03 uses a line stored in our pre-defined STRING structure (Chapter 5):

Function 9.03

```
/* Function line_length : Return length of line in structure line */
#include "cart_obj.h"
float line_length (line)
struct STRING line;
{
float distance(), sigma = 0.0; int i;
for (i = 1; i < line.number_of_points;i++)
     sigma+=distance(line.point[i].x,line.point[i].y,
          line.point[i-1].x,line.point[i-1].y);
return (sigma);
}
```

Notice that if the line has *line.number_of_points* points, we have *line.number_of_points*−1 segments.

Centroids

A transformation of point data which yields a point rather than a scalar is the computation of the centroid. Let's say that every point in the United States has a location, (x,y), and also has a population. A fairly simple concept is the center of gravity of the population. This single point is representative of a whole scatter of other points, so this is a points-to-point transformation. The Census Bureau usually publishes maps showing how this centroid has moved over time in the United States. How do we calculate this average or descriptive location? Locations in geographic space, perhaps the United States, have eastings (x), northings (y), and values, perhaps populations P. Each instance of the data will be called the "i"th, and there are a total of "npts" of them. Assuming planar geometry, we can calculate the average x, $\bar{x}$, the average y, $\bar{y}$, and weight them by their populations.

$$\bar{x} = \frac{\sum_{i=1}^{i=npts} P_i x_i}{\sum_{i=1}^{i=npts} P_i}$$

9.09a

$$\overline{y} = \frac{\sum\limits_{i=1}^{i=npts} P_i y_i}{\sum\limits_{i=1}^{i=npts} P_i} \qquad\qquad 9.09b$$

For example, if a city has a population of 12,000, we treat this as 12,000 cities at the same place with a population of only one person.

Standard Distance

Above we computed the mean x and y as an aggregate measure of location, so similarly we can compute the standard deviations in x and y as a measure of dispersal. The standard deviations in x and y can be used to compute the standard distance, a measure of the degree of scattering of point distributions around a focal point, the mean center.

$$S_x = \frac{\sum\limits_{i=1}^{i=npts} (x_i - \overline{x})^2}{npts} \qquad\qquad 9.10a$$

$$S_y = \frac{\sum\limits_{i=1}^{i=npts} (y_i - \overline{y})^2}{npts} \qquad\qquad 9.10b$$

The root of the sums of these values squared gives a value called the standard distance.

$$standard\_distance = \sqrt{S_x^2 + S_y^2} \qquad\qquad 9.11$$

This value is a measure of dispersal for point distributions. Now we have a simple measure of location and dispersal for point distributions. This value can be used analytically, for example to examine the assumptions surrounding interpolation to a grid.

Nearest-Neighbor Statistic

Another descriptive statistic of point distributions is the nearest-neighbor statistic (NNS). For npts points (x,y) scattered over an area A,

$$NNS = 2\,\frac{\sum\limits_{i=1}^{npts} d_i}{npts\,\sqrt{(A/npts)}} \qquad\qquad 9.12$$

where d is the distance from each point to its closest neighbor. In this case, the point distribution has an empirically derived scalar statistic which actually measures the resemblance of the point distribution to a clustered, a random, a grid, or a hexagonal point distribution, although the value is sensitive to the area used.

Function 9.04

```
/*
 * Function nearest : Return nearest neighbor statistic assumes
 * data are in structure POINT. Calling function should provide
 * area. kcc 10-88
 */
#include "cart_obj.h"
#include <math.h>
float    nearest(n_point, point, area)
int    n_point; struct POINT  point[MAXPTS]; float   area;
{
    float           distance(), this_distance, shortest_distance,
        sigma = 0.0;
    int             i, j;
    for (i = 1; i < n_point; i++) {
        shortest_distance = distance(point[0].x, point[0].y,
            point[i].x, point[i].y);
    for (j = 0; j < n_point; j++) {
        if (i != j) {
            this_distance = distance(point[i].x, point[i].y,
                point[j].x, point[j].y);
            if (this_distance < shortest_distance)
                shortest_distance = this_distance;
        }
    }
    sigma += shortest_distance;
    }
    return ((2.0 *
        sigma / (n_point * sqrt(area / (double) n_point)))));
}
```

9.3.2 Transformations Based on Lines

Intersection Point of Two Lines

Finding the intersection point of two lines is an example of an algorithm which performs a cartographic transformation yielding a point. A critical use of this algorithm in analytical and computer cartography is in clipping and in computing the overlap between sets of chain-encoded areas or the intersection of line features. This algorithm falls into the category of a test, since in computer cartography we frequently have to test lines for intersection. If two lines intersect, where do they intersect? This computation of line intersection points is an essential part of many vector mode systems, and lies at the basis of all clipping algorithms. For two line segments consisting of (x_1,y_1) to (x_2, y_2) and (x_3,y_3) to (x_4, y_4) we wish to compute the intersection point (p, q).

We can find the intersection point by solving two simultaneous equations for the line $y = a + bx$, with two known values. If (x_1,y_1) and (x_2,y_2) lie on the same line, then

$$y_1 = a_1 + b_1 x_1 \tag{9.13a}$$

$$y_2 = a_1 + b_1 x_2 \tag{9.13b}$$

Similarly, if (x_3,y_3) and (x_4,y_4) lie on the same line, then

$$y_3 = a_2 + b_2 x_3 \tag{9.13c}$$

$$y_4 = a_2 + b_2 x_4 \tag{9.13d}$$

If there exists an intersection point, (p, q) which lies on both lines, then

$$q = a_1 + b_1 p \tag{9.14a}$$

$$q = a_2 + b_2 p \tag{9.14b}$$

By subtracting the former from the latter,

$$q - q = a_2 - a_1 + p(b_2 - b_1) \tag{9.15a}$$

and rearranging, we obtain

$$a_1 - a_2 = p(b_2 - b_1) \tag{9.15b}$$

This implies that

$$p = \frac{a_1 - a_2}{b_2 - b_1} \tag{9.16}$$

and by substitution

$$q = a_1 + b_1 \frac{a_1 - a_2}{b_2 - b_1} \tag{9.17}$$

The final equation is worth a closer look. It will always provide a value, except in the case when the gradients (b) of the two lines are equal, in which case the fraction goes to infinity. This is the case of parallel lines, which includes the cases of coincident and collinear lines. Second, the point (p, q) may not lie on the line segments in question. It may lie on the extension of the segment beyond its end points. To test for this, we merely have to test to see if p lies between x_1 and x_2, and q lies between y_1 and y_2. If the coincidence is close to exact, the intersection is at one of the end points. Function 9.05 is designed to implement this algorithm.

Function 9.05

```
/* intersect: Find the point of intersection between two lines, if
/* it exists returns 0 for no intersection, else 1.        kcc 10-88
*/
#include "cart_obj.h"
#define SMALL 0.000001
int intersect(l1, l2, p)
     struct LINE_SEGMENT l1, l2; struct POINT     *p;
{
     int               vertical_case = 0, collinear_case = 0;
     float             xdif, a1, a2, b1, b2, hi1x, hi2x, hi1y, hi2y,
          lo1x, lo2x, lo1y, lo2y;
     xdif = l1.end.x - l1.begin.x;
     if ((xdif * xdif) < SMALL) vertical_case = 1;
     else { b1 = (l1.end.y - l1.begin.y) / xdif;
          a1 = l1.begin.y - (b1 * l1.begin.x); }
     xdif = l2.end.x - l2.begin.x;
     if ((xdif * xdif) < SMALL) vertical_case += 2;
     else { b2 = (l2.end.y - l2.begin.y) / xdif;
          a2 = l2.begin.y - (b2 * l2.begin.x); }
     switch (vertical_case) {
     case 0:
          /* Neither link is vertical */
          if ((b1 - b2) >= SMALL) { p->x = (a2 - a1) / (b1 - b2);
               p->y = a1 + (b1 * p->x);
          } else { /* The two links have equal slopes */
               if ((a1 - a2) >= SMALL) /* Collinear links */
                    collinear_case = 1;
               /* The links are parallel and not collinear */
               else return (0); }
          break;
     case 1:
          /* First link is vertical */
          p->x = l1.begin.x; p->y = a2 + b2 * p->x; break;
     case 2:
          /* Second link is vertical */
          p->x = l2.begin.x; p->y = a1 + b1 * p->x; break;
     case 3:
          /* Both lines are vertical */
          if ((l1.begin.x - l2.begin.x) >= SMALL) return (0);
          else /* Lines are vertical and collinear */
               collinear_case = 1; break;
     default:
          printf("Error, links do not match case set\n"); return (-1);
     }
     /* Store x ranges of first link */
```

```
hi1x = l1.end.x; lo1x = l1.begin.x;
if (l1.end.x < l1.begin.x)
    { hi1x = l1.begin.x; lo1x = l1.end.x; }
/* Store x ranges of second link */
hi2x = l2.end.x; lo2x = l2.begin.x;
if (l2.end.x < l2.begin.x) {
    hi2x = l2.begin.x; lo2x = l2.end.x; }
/* Store y ranges of first link */
hi1y = l1.end.y; lo1y = l1.begin.y;
if (l1.end.y < l1.begin.x)
    { hi1y = l1.begin.y; lo1y = l1.end.y; }
/* Store y ranges of second link */
hi2y = l2.end.y; lo2y = l2.begin.y;
if (l2.end.y < l2.begin.y)
    { hi2y = l2.begin.y; lo2y = l2.end.y; }
if (collinear_case) { p->y = lo2y; p->x = lo2x; }
/* Test to see if intersection point falls on both links */
if ((p->x <= hi1x) && (p->x >= lo1x) && (p->y <= hi1y) &&
    (p->y >= lo2y) &&
  (p->x <= hi2x) && (p->x >= lo2x) && (p->y <= hi2y) &&
    (p->y >= lo2y))  return (1);  else return (0);
}
```

Note that this function uses two different structures, LINK and POINT. A LINK is a one-dimensional cartographic object connecting two nodes. The POINT structure has been used previously in this chapter. Both structures were introduced in Chapter 5, and are derived from the digital cartographic data standards. Finally, it should be noted that the intersection algorithm depends upon a value SMALL, which is a tolerance. Setting SMALL to some arbitrary value is not a good idea, and repeated use of the algorithm with different values is suggested as a way to understand the relationship between scale and precision. An alternative tolerance could be one half of the length of the shortest line segment, or the step size of the resolution determined from the data quality report.

Distance of a Point to a Line

Yet another critical point transformation in automated cartography is establishing the relationship between a point and higher-dimension objects, such as lines and areas. To establish the distance between a point and a line, it is necessary to determine which link or line segment within the line is closest to the point. This can be done by using the point-to-point distance algorithm above, and testing each intermediate vertex of the line for distance.

Next, the closest links (and there may be more than one, depending on the line) should be tested to determine the closest point along the line to the test point. An easy way to find the intersection point on the line is to mirror the point about the line, and then to use the line intersection algorithm above to yield the point. Note that the result may be a point which is not on the line segment in question, in which case it should be discarded.

No program for this algorithm is presented here, since the more common carto-graphic problem is to generate a second line which is at some distance from the first, the so-called line-buffer problem (Schwarz, 1986).

9.3.3 Transformations Based On Areas

Area of a Polygon

How do we take a closed line (or POLYGON) and calculate area? This is an area to a scalar transformation. Say that we have a closed polygon and its coordinates are (x_1,y_1) to (x_n,y_n) and we repeat (x_1,y_1) to close the polygon, Let's give the formula and then prove empirically that it works.

$$area = \frac{1}{2} \left| \sum_{i=1}^{i=npts+1} [(x_i y_{(i-1)}) - (x_{(i-1)} y_i)] \right| \qquad 9.18$$

Function 9.06 is a computer program to compute the area of a closed polygon stored in our pre-defined POLYGON structure (Chapter 5). The returned value will be negative if the polygon is defined counter-clockwise.

Function 9.06

```
/* Function polygon_area: Return area of polygon in structure
      polygon */
#include "cart_obj.h"
float area_of_polygon  (polygon)
struct POLYGON polygon;
{
    float sigma = 0.0; int i;
    for (i = 1; i < polygon.boundary.number_of_points; i++){
        sigma += (polygon.boundary.point[i].x *
             polygon.boundary.point[i-1].y) -
             (polygon.boundary.point[i-1].x *
             polygon.boundary.point[i].y);
    return (0.5 * sigma);
}
```

Figure 9.06 and Table 9.1 show a simple polygon and the computation of the area by two methods, an intuitive proof and use of the equation. For simplicity, let

$$A = x_i y_{(i-1)}$$

and

$$B = x_{(i-1)} y_i$$

TABLE 9.1   AREA OF A POLYGON

Vertex	x	y	A	B	Difference
1	1	1	-	-	-
2	1	6	1	6	-5
3	3	6	18	6	12
4	3	5	18	15	3
5	2	5	10	15	-5
6	2	4	10	8	2
7	3	4	12	8	4
8	3	3	12	9	3
9	2	3	6	9	-3
10	2	2	6	4	2
11	3	2	6	4	2
12	3	1	6	3	3
13	1	1	1	3	-2

					16

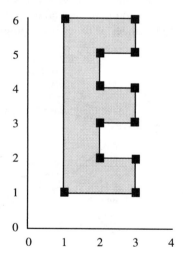

Figure 9.06   Sample Polygon for the Worked Example

Notice that it is easy to accumulate area as we go, so the computational values can be accumulated while we digitize a polygon.

By simple inspection of Figure 9.06 we can see that the polygon consists of a letter "E". This letter can be made up from eight unit-square building blocks, five forming a vertical column, and three attached to the right in alternate rows. Clearly, by inspection, the area of the polygon is eight units. From the table we can see that the sum of the difference column is 16. If we take half of this value, as required in the formula, we get the correct area of eight units. Also note that the value of the area comes out negative if the polygon is specified counter-clockwise. One way we can use this is to include holes within a polygon as rings within the polygon, except specified in the other direction. Then to find the total area, we simply add the area to the area of its enclosed holes to get the actual area of the polygon. Holes within the polygon under the digital cartographic data standards are considered as rings, with no retraced lines.

Point-in-Polygon Test

The next-highest-dimension object for which point relations can be established is the area or polygon. There are many algorithms for testing whether or not a point falls within a polygon. Many of the algorithms use area computations, and many others work only for convex polygons, or polygons without concavities. These algorithms are virtually useless in analytical cartography, for the shapes of real-world cartographic objects are far from geometric, and almost never entirely convex in their boundaries. Probably the best and the simplest algorithm for point-in-polygon testing is the Jourdan arc theorem. This simply states that a line between a point known to be outside a polygon will cross the polygon boundary an even number of times if the point is outside the polygon and an odd number of times if the point is inside the polygon (Figure 9.07).

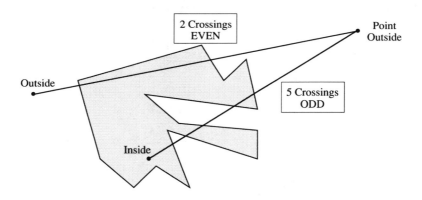

Figure 9.07   The Jourdan Arc Theorem for Point-in-Polygon Testing

This theorem is remarkably robust, except when the lines either touch a vertex or run parallel to an edge. The advanced student may wish to consider how the following program and the line intersection algorithm above can be modified to test correctly for points which lie on the boundary. Function 9.07 below works only for what is defined in the standards as the interior area of the polygon. It establishes a point which is outside the polygon by finding the maximum x value for the polygon and multiplying it by 2.

Function 9.07

```
/* Function to perform point in polygon test using Jourdan arc
      theorem    kcc 5-88 */
#include "cart_obj.h"
int point_in_polygon (point, polygon)
struct POINT point;       struct POLYGON polygon;
{
struct POINT pt; struct LINK link1, link2;
int i, intersections = 0;
link1[0].x = point.x; link1[0].y = point.y;
/* Establish a point outside the polygon */
link1[1].y = polygon.boundary.point[0].y;
      link1[1].x = polygon.boundary.point[0].x;
for (i=1; i < polygon.boundary.number_of_points;i++)
      if (polygon.boundary.point[i].x > link[1].x)
            link[1].x = polygon.boundary.point[i].x;
link[1].x *= 2.0;
/* Now check the link from the test point to the outside */
/* point against each of the polygon's boundary segments */
      for (i=1; i < polygon.boundary.number_of_points;i++) {
      link2[0].x = polygon.boundary.point[i-1].x;
      link2[0].y = polygon.boundary.point[i-1].y;
      link2[1].x = polygon.boundary.point[i].x;
      link2[1].y = polygon.boundary.point[i].y;
      if (intersect(link1, link2, pt)) intersections++;
      }
/* Return odd/even, 1 means inside, 0 outside */
return (intersections % 2);
}
```

Thiessen Polygons

The final transformation between a point and a higher-dimension object is a point-to-polygon transformation. The earliest computer cartographic program to perform this transformation was the SYMAP package, which called the polygons created, *proximal regions*. These regions are called *Thiessen polygons* after the climatologist A. H. Thiessen. In geography, the term *Voronoi diagram* is used. Thiessen devised these regions to assist in the interpolation of climatic data from unevenly distributed weather stations. The method is used wherever data have been collected at points and the cartographer desires to use area-based analytical techniques or symbolization methods. As such, this is a space-partitioning transformation, since it divides the regions surrounding a set of points exclusively into a set of polygons. The algorithms for finding the Thiessen polygons is too lengthy to be included here; the reader is referred to Brassel and Reif's paper (1979), which discusses a FORTRAN implementation, and to Mark's (1987) more recent paper, which uses quad trees as a transformational tool to determine the regions. One noteworthy side effect of computing these areas is that the Delaunay triangulation is also performed since it is the dual of the Voronoi diagram. This transformation partitions the space using the points as nodes in the network, and produces a TIN (triangulated irregular network) structure.

9.4 AFFINE TRANSFORMATIONS

Map projections are the only coordinate transformations covered so far. It is possible, if we assume plane geometry, to provide a set of universal transformations for conversion between any two Euclidean coordinate spaces. These transformations, known as *affine* transformations, are actually three distinct transformations applied in sequence. The sequence is first, a *translation*, in which the coordinate system origin is moved to a new location, second, a *rotation*, in which the axes of the coordinate system are rotated to match the new system, and finally, a *scaling*, in which the numbers along the axes are scaled to represent the new space scale. It is possible to accumulate the effects of these three transformations into a single step, which can then be applied on a feature-by-feature basis.

First, a transformation to move the origin can be stated as

$$\begin{bmatrix} x & y & 1 \end{bmatrix} \begin{bmatrix} 1 & 0 & 0 \\ 0 & 1 & 0 \\ -x_0 & -y_0 & 1 \end{bmatrix} = \begin{bmatrix} x - x_0 & y - y_0 & 1 \end{bmatrix} \qquad 9.19$$

Note the similarity between the translation and the movement of the origin for a map projection. Map projections usually start by subtracting from the longitude the longitude at which the projection is to be centered, equivalent to rotating the earth to a new center of view.

The next stage is a rotation of the axes about an angle θ.

$$\begin{bmatrix} x - x_0 & y - y_0 & 1 \end{bmatrix} \begin{bmatrix} \cos\theta & \sin\theta & 0 \\ -\sin\theta & \cos\theta & 0 \\ 0 & 0 & 1 \end{bmatrix} = \qquad 9.20$$

$$\begin{bmatrix} \cos\theta(x - x_0) - \sin\theta(y - y_0) & \sin\theta(x - x_0) + \cos\theta(y - y_0) & 1 \end{bmatrix}$$

Third, the point is subject to a scaling of the axes:

$$\begin{bmatrix} \cos\theta(x - x_0) - \sin\theta(y - y_0) & \sin\theta(x - x_0) + \cos\theta(y - y_0) & 1 \end{bmatrix} \begin{bmatrix} S_x & 0 & 0 \\ 0 & S_y & 0 \\ 0 & 0 & 1 \end{bmatrix} \quad 9.21$$

$$= \begin{bmatrix} S_x(\cos\theta(x - x_0) - \sin\theta(y - y_0)) & S_y(\sin\theta(x - x_0) + \cos\theta(y - y_0)) & 1 \end{bmatrix}$$

Note that the scaling looks much like the scale transform in a map projection. This is precisely what the map projection transformation above is doing. The sequence of these transformations is translation, rotation, and scaling. Often order is important; in other words, if we use some other sequence, we will not end up with the same final location. That has implications for inverting this transformation. Since the transformation is multi-step and we have to perform the steps in the correct sequence, we also have to undo them in the correct sequence.

Figure 9.08 shows these three transformations graphically. We can multiply these 3 x 3 matrices by each other and produce a single transformation which summarizes the entire set of transformations.

$$\begin{bmatrix} x & y & 1 \end{bmatrix} \begin{bmatrix} T \end{bmatrix} \begin{bmatrix} R \end{bmatrix} \begin{bmatrix} S \end{bmatrix} = \begin{bmatrix} x' & y' & 1 \end{bmatrix} \qquad 9.22$$

TRANSLATION   ROTATION   SCALING

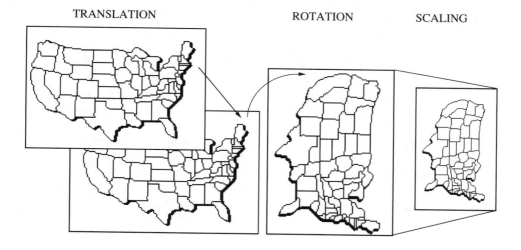

Figure 9.08   Affine Transformations

In non-matrix algebra, this yields:

$$x' = S_x(\cos\theta(x - x_0) - \sin\theta(y - y_0)) \qquad 9.23a$$

$$y' = S_y(\sin\theta(x - x_0) + \cos\theta(y - y_0)) \qquad 9.23b$$

Note that the x and y of the origin are the x and y of the new origin in the old coordinate system. Consider the following example. A point in the old coordinate system is at (2, 2). What is the coordinate of the point in a system which has a new origin at point (0, 2) in the old coordinate system, has a 30-degree clockwise rotation of the two sets of axes, and has new axes which have twice the spacing? Using the formulas

$$x' = 2(\cos 30^o(2 - 0) - \sin 30^o(2 - 2))$$

$$y' = 2(\sin 30^o(2 - 0) + \cos 30^o(2 - 2))$$

This reduces to

$$x' = 2(0.866 * (2) + 0) = 3.464$$
$$y' = 2(1 + 0) = 2$$

9.5 STATISTICAL SPACE TRANSFORMATIONS

9.5.1 Rubber Sheeting

In many cases, including those in which direct measurement of a cartographic entity has taken place, the precise space geometry is unknown. This is especially the case with remotely sensed images and air photos, when the viewing geometry is unknown. In both cases, the image is distorted for a number of reasons, including the flight geometry, the earth's curvature, the portion of the earth covered, the view angle, the topography, the altitude of the observation platform, the characteristics of the lens, and a host of other reasons. Also, the atmosphere itself is far from uniform, making its own "lens" effects, depending on pressure, moisture content, time of day, solar heating, etc. Even under the best of circumstances, aircraft, and even satellites, pitch and yaw, and even "wobble". The result, the earth's surface as seen in an image, is an approximate map, in which the actual cartographic objects and features can be plainly seen, but the precise relationships between scale, area, shape, and direction are "fuzzy".

An alternative way to state the relationship between such a distorted image and an accurate cartographic map base is to state that the precise map projection transformation is unknown. Instead, we have numerous actual examples of the transformation having been applied. To fully describe the projection characteristics, necessary if we wish to "undo" its effects and convert the image back to actual map space, we have to study the projection empirically.

An analogy is to imagine the map printed on a rubber sheet. When the air photo or satellite image arrives, we should be able to recognize precisely the locations of a key set of places on the image. These locations are the *ground control points*. A ground control point is a physical feature in a scene which is detectable, and whose precise coordinates in a coordinate system are known. For some purposes, merely the location of ground control points is enough, but for others we should know the location and elevation quite precisely and with a high degree of accuracy. Bernstein (1983) noted that typical ground control points are airports, highway intersections, land-water interfaces, and geological and field patterns. Campbell (1987) also noted that the ground control points should ideally be a single pixel in size on the image, and should be easily identifiable against their background. They should be dispersed throughout the image, particularly at the edges, not all concentrated in a single region.

To follow the analogy, the ground control points can be seen as holes, through which the rubber sheet correct map is pinned to the image such that the map is stretched to fit correctly, i.e., the pins are at their correct locations on both the map and the image. Given this, we can measure the "stresses" in the rubber sheet to figure out how to stretch it back to its initial state. We then apply these "stresses" to the image, and as a result the entire image is stretched into the map space.

Mathematically, what actually happens is that we have sets of points which match in the two spaces. These points have n locations (x,y) on the map and (u,v) in the image.

$$\begin{bmatrix} x \\ y \end{bmatrix} = \begin{bmatrix} T \end{bmatrix} \begin{bmatrix} u \\ v \end{bmatrix} \qquad 9.24$$

The trick is to use the n multiple cases of (x,y) and (u,v), for which we have ground control points, to estimate T. The process of estimation proceeds by using what Sprinsky (1987) called the six-parameter affine, i.e.,

$$u = \beta_0 x + \beta_1 y + \beta_2 \qquad 9.25a$$

$$v = \beta_3 x + \beta_4 y + \beta_5 \qquad 9.25b$$

The beta coefficients can be computed by a sort of least squares fit. If we let:

$$p = \frac{\sum_{i=1}^{i=n} \left[(x_i - \bar{x})(y_i - \bar{y}) \right]}{\sum_{i=1}^{i=n} (y_i - \bar{y})^2} \qquad 9.26$$

and

$$q = \frac{\sum_{i=1}^{i=n} \left[(x_i - \bar{x})(y_i - \bar{y}) \right]}{\sum_{i=1}^{i=n} (x_i - \bar{x})^2} \qquad 9.27$$

then the following sets of equations allow the solution of the six unknown betas from n control points.

$$\beta_3 = \sum_{i=1}^{i=n} \left[(x_i - \bar{x})(v_i - \bar{y}) \right] - p\frac{\sum_{i=1}^{i=n} \left[(y_i - \bar{y})(v_i - y_i) \right]}{\sum_{i=1}^{i=n} (x_i - \bar{x})^2} - p\sum_{i=1}^{i=n} \left[(y_i - \bar{y})(x_i - \bar{x}) \right] \qquad 9.28$$

$$\beta_1 = \sum_{i=1}^{i=n} \left[(y_i - \bar{y})(u_i - x_i) \right] - q\frac{\sum_{i=1}^{i=n} \left[(x_i - \bar{x})(u_i - x_i) \right]}{\sum_{i=1}^{i=n} (y_i - \bar{y})^2} - q\sum_{i=1}^{i=n} \left[(x_i - \bar{x})(y_i - \bar{y}) \right] \qquad 9.29$$

$$\beta_4 = 1 - p\beta_3 + \frac{\sum_{i=1}^{i=n} \left[(y_i - \bar{y})(v_i - y_i) \right]}{\sum_{i=1}^{i=n} (y_i - \bar{y})^2} \qquad 9.30$$

$$\beta_0 = 1 - q\beta_1 + \frac{\sum\limits_{i=1}^{i=n}\left[(x_i - \overline{x})(u_i - x_i)\right]}{\sum\limits_{i=1}^{i=n}(x_i - \overline{x})^2} \qquad\qquad 9.31$$

$$\beta_2 = \frac{1}{n}\sum_{i=1}^{i=n}(u_i - x_i) + \overline{x} - \beta_0\overline{x} - \beta_1\overline{y} \qquad\qquad 9.32$$

$$\beta_5 = \frac{1}{n}\sum_{i=1}^{i=n}(v_i - y_i) + \overline{y} - \beta_4\overline{y} - \beta_3\overline{x} \qquad\qquad 9.33$$

The inverses of these formulas can now be used to allow the image coordinates as input, and get map coordinates as output.

$$x = \alpha_0 u + \alpha_1 v + \alpha_2 \qquad\qquad 9.34a$$

$$y = \alpha_3 u + \alpha_4 v + \alpha_5 \qquad\qquad 9.34b$$

where

$$\alpha_0 = \frac{\beta_4}{\beta_4\beta_0 - \beta_3\beta_1} \qquad\qquad 9.35$$

$$\alpha_1 = -\frac{\beta_1}{\beta_4\beta_0 - \beta_3\beta_1} \qquad\qquad 9.36$$

$$\alpha_2 = \frac{\beta_1\beta_5 - \beta_2\beta_4}{\beta_4\beta_0 - \beta_3\beta_1} \qquad\qquad 9.37$$

$$\alpha_3 = -\frac{\beta_3}{\beta_4\beta_0 - \beta_3\beta_1} \qquad\qquad 9.38$$

$$\alpha_4 = \frac{\beta_0}{\beta_4\beta_0 - \beta_3\beta_1} \qquad\qquad 9.39$$

$$\alpha_5 = \frac{\beta_3\beta_2 - \beta_5\beta_0}{\beta_4\beta_0 - \beta_3\beta_1} \qquad\qquad 9.40$$

This solution is also frequently used in surveying to adjust errors in measurement over a set of readings, and is frequently developed as a matrix solution (Moffitt and Bouchard, 1982). Several of the terms in the solution above are clearly determinants and transposes. Once more, we have an example of a cartographic transformation. However, in this case we have had to deduce the exact nature of the transformation from its results, and have inverted the transformation empirically, in a "best-fit" solution. When there is no control over the transformation, or when the transformation is very complex, this is the only approach left. Finally, it should be noted that in remote sensing the final result is a grid of points which are not regular. The points are therefore immediately resampled back into a regular grid which is orthogonal to the map coordinate system and has some finite pixel size. Methods for performing this second step transformation are discussed in Bernstein (1983).

9.5.2 Cartograms

Cartograms, also known as value-by-area maps, varivalent projections, and pycnomirastic projections (Tobler, 1986), are maps drawn so that the areas of their internal enumeration units are proportional to the units they represent (Dent, 1985, p. 326). Generally, they are of two types. *Non-contiguous* cartograms simply shrink the scale factor of an areal unit in isolation from all others, and present the results as a diagram in which geographic space is interrupted. As such, these maps could be regarded as a form of proportional point symbol mapping. *Contiguous* cartograms maintain the continuity of geographic space, and as such are really a variation upon map projections. Like all map projections, direction, distance, shape, area, and the graticule are distorted in combination on these maps. Their advantage is that the map base can be distorted to make the mapping of a thematic data item more appropriate, removing the underlying differences in population density when showing per capita income, for example.

Tobler (1986) advocated the following method for the computation of "pseudo-cartograms". First, the data to be used to "reshape" the map should be converted to the structure of a uniform grid, with equal latitude-longitude spacing north-south and east-west. If this grid is an m by n array, **Z,** then it can be written as the sum of k arrays, where k is the lesser of m and n. The arrays can then be written as a set of products of each of two k by 1 vectors, **U** and **V.**

$$Z = U_1 V_1 + U_2 V_2 + ... + U_k V_k \qquad 9.41$$

where

$$U_k = k \ th \quad eigenvector \ of \ Z \ Z^T \qquad 9.42$$

and

$$V_k = \frac{Z^T U_k}{U^T U_k} \qquad 9.43$$

The vectors can then be sorted by the amount of variance accounted for by the model, and the first picked to yield the transforms. The **U** vectors are now functions only of the row index or y, and the **V** vectors are functions of the column index or x. These values are then projected using a standard equal-area map projection.

$$y = R \ s \ U_1 \ cos(y) \qquad 9.44$$

$$x = R \ s \ V_1 \qquad 9.45$$

Tobler has published extensively on these transformations, including a set of computer programs (Tobler, 1974). An example of a cartogram based on population density is shown in Figure 9.09.

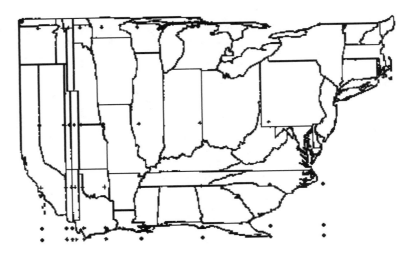

Figure 9.09   A Pseudo-cartogram of U.S. Population (from Tobler, 1986, p. 49)

9.6 SYMBOLIZATION TRANSFORMATIONS

As far as producing a map is concerned, the final transformation is the symbolization transformation. In this transformation, the cartographic data is subjected to all the necessary transformations to use a specific map type, and the map itself is generated. While the actual look of the map is determined largely by the type of data, the fundamental properties of the data, the type of map, and the design of the map, there are some common or generic symbolization transformations. Among these is the viewing transformation, in which the geometry of the map, rather than the geometry of the data is established. Also important is the actual plotting of the graphic elements, how they are symbolized, and how the text is added. This section contains information about these final, often critical, symbolization transformations according to the GKS standard.

9.6.1 The Normalization or Viewing Transformation

A point-to-point transformation of great importance in analytical and computer cartography is the transformation between the *world* coordinates, given in the coordinate system required by an application, and the coordinates of a particular display device, such as the screen of a graphics terminal. Typical *world* coordinate systems are either arbitrary, map millimeters for example, or standardized, such as UTM coordinates. The display device also uses arbitrary coordinates known as *device* coordinates usually the number of pixels which are addressable in the east-west and north-south directions. Since each particular display device has a different number of addressable pixels, it is desirable to have an intermediate coordinate system with a standard size into which we can transform the map. The intermediate coordinates are known as *normalized device* coordinates. In the language of the graphical kernel system, the transformation from the world to the normalized device coordinates is known as the *normalization* transformation, while the transformation to device or screen coordinates is the *workstation* transformation. These transformations are summarized in Figure 9.10.

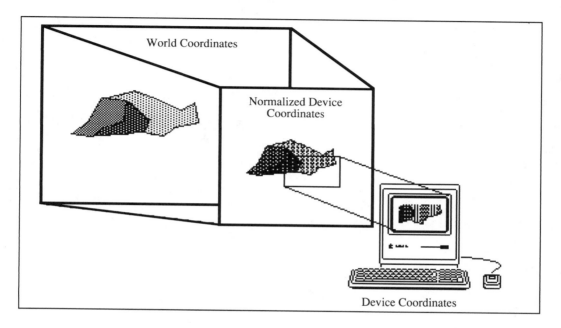

Figure 9.10   Normalization and Workstation Transformations

The normalization transformation is specified by defining the limits of an area in the world coordinate system known as the *world window*. This rectangle or square is mapped onto a specified part of the normalized device coordinate space called the *world viewport*. Window and viewport limits are assumed to be parallel to the coordinate axes, so any rotations must be handled by the user, using the rotation algorithms given above. Normalized device coordinates range from (0.0,0.0) at the lower left corner to (1.0,1.0) at the upper right.

Similarly, we can control how much of the normalized device coordinate space is visible by establishing a workstation transformation. This is done by specifying a workstation window in normalized device coordinates. The actual transformations, the mathematics of which can be derived from the transformational equations given above, are contained within the graphical kernel system calls *set_window* and *set_viewport*. A small interactive program to establish the transformation is given as Function 9.08. In this example, the goal is to center the map onto the largest possible piece of the device space, while maintaining the correct scale relationships between x and y. This is done with the help of the GKS inquiry function *inquire_display_space_size*. This function returns the device coordinate units, the size of the surface of the device, and the number of pixels the device is capable of displaying.

Function 9.08

```
/* Establish the normalization transformation for an active display
/* device, using largest available centered rectangle with
/* appropriate scale. kcc    5-88
*/
#include "gks.h"
normalize() {
    float llx, lly, urx, ury, xmin, xmax, ymin, ymax;
    float units, xpix, ypix, width, height, aspect_ratio, map_size;
    /* Establish the size of the display */
    inquire_display_surface_size (&units, &xmax, &ymax, &xpix, &ypix);
    printf(" Enter the world window\n");
    printf(" Enter (xmin, ymin) :"); scanf ("%f%f%*1c", &llx, &lly);
    printf(" Enter (xmax, ymax) :"); scanf ("%f%f%*1c", &urx, &ury);
    /* Establish the aspect ratio of the world window */
    width = urx - llx; height = ury - lly;
        aspect_ratio = height / width;
    /* Apply the same aspect ratio to the display device */
    if (aspect_ratio < 1.0) { /* Landscape image */
        map_size = xmax * aspect_ratio;
        ymin = (ymax - map_size) / (2.0 * xmax);
        ymax = (ymin + map_size) / xmax; xmax = 1.0;
        }
    else { /* Portrait Image */
        map_size = ymax / aspect_ratio;
        xmin = (xmax - map_size) /( 2.0 * ymax);
        xmax = (xmin + map_size) / ymax;
        ymax = 1.0 }
    /* Finally set up the normalization transformation */
    set_window ( llx, urx, lly, ury);
    set_viewport ( xmin, xmax, ymin, ymax);
}
```

9.6.2 Primitives and Attributes

Under GKS, one of the basic tasks of the system is to generate pictures. In the case of computer cartography, the pictures contain symbolizations of cartographic objects i.e., maps. In the terminology of the digital cartographic data standards, the real-world phenomenon to be represented on a map is called a cartographic entity. The digital representation, usually as a data structure, of the entity is as a cartographic object. In the final cartographic transformation, the *symbolization* transformation, the cartographic object is depicted as a cartographic element, i.e., a distinct part of the graphic of a map.

As such, the cartographic map element is a picture under GKS terminology, and therefore consists of a collection of *primitives* with their associated *attributes*.

GKS defines and makes available to the programmer six primitives for digital drawing. These consist of one line primitive, one point primitive, one text primitive, two raster primitives, and one general purpose primitive. The final cartographic transformation is to convert the cartographic data into groups of these primitives, and to set their attributes.

The line primitive is called the *polyline*. In symbolizing a polyline, GKS generates a set of straight-line segments connecting a given point sequence. The point primitive is the *polymarker.* 'To symbolize a polymarker, GKS generates a symbol of a given type at a specified location. Symbols can be points, circles, squares, etc. The text primitive in GKS is *text*. Text causes GKS to generate a text character string at a specified location.

The two raster primitives under GKS are *fill area* and *cell array*. Fill area is used to generate polygons, and to control the symbolization of their interiors. Cell array generates an array or grid of rectangular cells, each with individual colors. The cell array is an abstraction of the array of pixels on a raster type device. Finally, the *generalized drawing primitive* is designed to make use of special hardware characteristics of a specific workstation. Some workstations can generate circles, ellipses, spline curves, etc. These primitives have the workstation geometric transformations applied to them, but actual plotting is left up to the workstation.

Attributes are a function of the primitive to which they apply. Generally, polylines have the attributes of *line width, line type*, and *line color*. Polymarkers have the attributes of *marker size scale factor, marker type*, and *color*. Text has by far the most attributes, including *character height, character up vector, character expansion factor, text path, character spacing, text font, text precision, color*, and *text alignment*. Fill areas have the attributes *pattern size, pattern reference point, pattern array, interior style, hatch style*, and *color*. The cell array has only *color*, while the attributes of the generalized drawing primitive are dependent upon its type. Another important attribute is the *pick identifier*, which controls the way the user may interact with the particular primitive when it is displayed on the workstation.

Among the large variety of possible settings, line types may be solid, dashed, or dashed-dotted; text path may be left, right, up, or down; character alignment may be left aligned, centered, or right aligned to top, normal (middle), or bottom; and interior styles may be hollow, solid, pattern, or hatch. In addition, the attributes may be individual, in which case they apply to all the primitives that follow until they are changed, or bundled, in which case the attributes may be collected together as a group and applied collectively to one or more primitives. Such a collection of primitives is stored for an application in a bundle table. *Indexes* named for each primitive allow separate management of these bundle tables.

It should be noted that GKS implementations usually come with library files, which pre-define certain attribute parameters so that more intuitive names can be used. These definitions are often in a header file, such as gksdef.h, but the actual name depends upon the GKS implementation. The pre-processor names, in keeping with the

C standard, are in uppercase. So, for example, the first eight colors, which require calls with integers 1 through 8 , are instead "#defined" to *RED, GREEN, BLUE, MAGENTA, CYAN, YELLOW, BLACK,* and *WHITE.*

9.6.3 Drawing Cartographic Objects

The very final stage in the cartographic transformation from data to map is to actually implement a primitive, after having set its attributes, either individually or in a bundle. Before the primitives can be accessed, a workstation must be open and the sets of geometric space transformations to be applied must be active. Assuming that this is the case, and that the computer cartographer has loaded data into the data structures we have outlined in earlier chapters, all that remains is to point to the GKS functions which actually do the display. First, we will cover the actual calls themselves, and following each function will be the relevant attributes which can be changed to manipulate the workstation display.

(1) *Polyline* To display a polyline, as stored in the previously defined STRING data structure, the GKS call is

```
polyline (string.number_of_points, &string.point[0].x,
    &string.point[0].y);
```

Attributes are

```
set_linewidth_scale_factor (scale);
set_linetype (type);
```

(2) *Polymarker* A polymarker is defined and stored in the previously defined POINT structure. The GKS drawing call is

```
polymarker (1, point.x, point.y);
```

Attributes are

```
set_marker_size_scale_factor (scale);
set_marker_type (type);
set_marker_color_index (color);
```

(3) *Text* Text has not been treated as having a cartographic data structure. In its simplest form, text consists of a string of C type *char*. The GKS drawing call is

```
text (text_string);
```

Attributes are

```
set_text_alignment (horizontal, vertical);
set_text_font_and_precision (font_index, precision_index);
set_text_path (path);
set_text_color_index (color);
```

(4) *Fill Area* To fill a polygon, the cartographer should use the previously defined

RING structure, into which has been loaded data. The GKS command to plot the polygon is

```
fill_area (ring.number_of_points, &ring.point[0].x,
        &ring.point[0].y);
```

Attributes are:

```
set_fillarea_color_index (index);
set_fillarea_interior_style (style);
set_pattern_reference_point (x, y);
set_pattern_size (size_x, size_y);
```

To fill a complex polygon, the polygon boundary should be filled first, followed by the holes.

(5) *Cell Array* To depict an array, such as a digital elevation model, or a hill-shaded grid on a workstation, the cartographer should use the previously defined GRID structure, into which has been loaded data. The GKS command to plot the entire grid cell map is:

```
grid_cell (x_p, y_p, x_q, y_q, no_cols, no_rows, col_start,
        row_start, dx, dy, &grid.z[0][0]);
```

The only attribute is

```
set_color_index (index);
```

(6) *Generalized Drawing Primitive* The depiction of a GDP is specific to a particular workstation, and is rarely used under cartographic applications unless drawing times become too slow.

9.7 SUMMARY

In this chapter we have covered numerous examples of cartographic transformations and have worked through specific examples of how this point of view allows the organization of the body of material which makes up computer cartography into its constituent parts within analytical cartography. Two major types of transformations have been presented. First, we have seen how cartographic objects can be transformed by dimension. As part of these dimensional transformations, the case of point-to-point transformations was used to show how the locational attributes of cartographic data can be transformed to produce map projections and cartograms. Finally, we have examined the symbolization transformation, in which the cartographic objects gain actual instances as GKS pictures, collections of graphics primitives with their associated attributes.

Transformations of the first type noted above are fully within the domain of analytical cartography. The symbolization transformation is the very essence of computer cartography, and we have made the natural step from discussion of algorithms directly to computer standards and functions. This programming theme forms part four of this book, *Producing the Map*. In the remainder of the current section, the following two chapters, we will look specifically at line and area transformations, at data structure transformations, and finally, at the transformations possible with volumetric data.

9.8 REFERENCES

BERNSTEIN, R. (Ed.) (1983) "Image geometry and rectification," in D. S. Simonett, (Ed.) *Manual of Remote Sensing* Volume I : *Theory, Instruments and Techniques,* American Society of Photogrammetry, Sheridan Press, Falls Church, VA, pp. 873-922.

BRASSEL, K. E., AND REIF, D. (1979) "A procedure to generate Thiessen polygons," *Geographical Analysis*, vol. 11, no. 3, pp. 289-303.

CAMPBELL, J. B. (1987) *Introduction to Remote Sensing*, Guilford Press, New York.

DENT, B. D. (1985) *Principles of Thematic Map Design*, Addison-Wesley, Reading, MA.

MARK, D. M. (1987) "Recursive algorithm for determination of proximal (Thiessen) polygons in any metric space," *Geographical Analysis*, vol. 19, no. 3, pp. 264-272.

MOFFITT, F. H., AND BOUCHARD, H. (1982) *Surveying*, Harper & Row, New York, Seventh Edition.

National Committee for Digital Cartographic Data Standards (1988) "The proposed standard for digital cartographic data," *The American Cartographer*, vol. 15, no. 1, pp. 21-28.

National Oceanic and Atmospheric Administration (1988), *The General Cartographic Transformation Package*, National Ocean Service, Charting and Geodetic Service, National Geodetic Survey, Rockville, MD.

SCHWARZ, C. R. (1986) "Algorithms for constructing lines separated by a fixed distance," *Proceedings of the Second International Symposium on Spatial Data Handling,* Seattle, WA.

SNYDER, P. (1983) *Map Projections Used by the U.S. Geological Survey*, Geological Survey Bulletin 1532, U.S. Government Printing Office, Washington, DC, Second Edition.

SNYDER, J. P. (1987) *Map Projections - A Working Manual*, Geological Survey Bulletin P-1395, U.S. Government Printing Office, Washington, DC.

SPRINSKY, W. H. (1987) "Transformation of positional geographic data from paper-based map products," *The American Cartographer*, vol. 14, no. 4, pp. 359-366.

TOBLER, W. R. (1974) *Cartogram Programs*, Department of Geography, University of Michigan, Ann Arbor.

TOBLER, W. R. (1979) "A transformational view of cartography," *The American Cartographer*, vol. 6, no. 2, pp. 101-106.

TOBLER, W. R. (1986) "Pseudo-cartograms," *The American Cartographer*, vol. 13, no. 1, pp. 43-50.

10

Map Data Structure Transformations

10.1 WHY TRANSFORM BETWEEN STRUCTURES?

In virtually all mapping applications it becomes necessary to convert from one cartographic data structure to another. The ability to perform these object-to-object transformations often is the single most critical determinant in a mapping system's flexibility. Why is this the case? A number of reasons dictates that the mapping process involve data structure conversions, and these are related to data input and geocoding, data storage and representation, the suitability of particular structures for different analytical and modeling demands, and finally, the demands of a particular symbolization transformation.

Data input has already been noted as a primary determinant of data structure. Particular data capture devices, such as scanners and semi-automatic digitizers, generate data in a specific form. Most frequently, the input determines how the particular cartographic entity becomes a cartographic object in digital form. Scanners, for example, generate grids, while semi-automated digitizers produce strings of (x,y) coordinates. Geocoding stamps a particular coordinate system, resolution, and map projection on the data. In virtually every case, the cartographer will find that the available digital cartographic data are in the wrong structure for the required type of cartographic object, are on the wrong map projection, have the wrong resolution for mapping, or the map rectangle needs to be rotated, scaled, or translated to produce the map. Analytical and computer cartography, unlike many other disciplines, has available large reserves of common, generically digitized cartographic data. As a result, cartographers must change data structures simply to move the data into the correct geographic extent and from an input format into the format required by a particular piece of mapping software.

In Chapter 5, consideration was given to the different mechanisms by which digital cartographic data can be stored within a computer's memory, and to how different means of data representation can save storage or improve data accessibility. Since one of the distinguishing characteristics of cartographic and geographic data is sheer volume, the conversion of data between structures, or between representations of a single structure, can save considerable storage space and processing time. The time and space limitations become more apparent as cartographic software moves from larger to smaller computers. While the amount of RAM and disk storage available to microcomputers has increased, and while mass storage technology is making significant breakthroughs in volume capabilities, the fact is that processing the millions of data points necessary for high-accuracy cartography really strains the microcomputer. Usually, precision or scale are sacrificed, with cartographically unacceptable results. Transforming data between cartographic data structures means that the application can optimize storage and processing time use as appropriate. Often the best data structure for a map depends on the demands placed upon the data for analytical or display purposes. Good cartographic software does not force data into a single structure, but retains the option to flip between structures on demand.

Analysis and modeling also require different data structures. As an example, the skeleton or medial axis transform of a polygon can be performed in both grid or polygon entity-by-entity mode. In entity-by-entity mode, the locus of the largest enclosed circles must be traced through the polygon. In grid mode, the polygons are simply eroded away step by step from the edges while connectivity is maintained, until the medial axis is formed. Each operation is fairly fast. Inverting the transformation in grid mode is simple. However, rebuilding the polygon from the medial axis is difficult in entity-by-entity mode. The grid data structure is suitable for modeling and analytical operations such as Fourier and principal component analysis, filtering, and edge detection. The TIN structure is good for modeling overland and stream hydrology, and for simulating erosion. Entity-by-entity definitions are good for high-precision output, with multiple-weight lines and elaborate shading. Continuous patterns are more suited to the grid. The relative advantages and disadvantages of the various structures are many, and even depend on the implementation characteristics, such as language, computer, and operating system.

For symbolization, actually producing the map, again certain structures meet different sets of cartographic demands, which implies that different types of map, different map scales, etc., all have their different optimal cartographic data structure. Often, the characteristics of the output device determine the best data structure. For example, raster devices favor grid and quad-tree structures, while plotters, laser-jet printers, and automatic scribers require data in line or polygon entity-by-entity format. The type of representation is also important. Choropleth maps can be produced in any structure, but hill-shading and perspective views are best using grids or TINs. As far as symbolization is concerned, nowhere is the influence of data structure more obvious than in map text labeling. Vector displays usually use the Hershey fonts, and are capable of some quite attractive precision lettering. Raster devices use bit-mapped graphics, and as such produce blocky text, which looks worse with enlargement, and is restrained in its angle on the screen. Just as many systems now support both raster and vector, the now common PostScript graphics merge vector draw commands with the raster capabilities of laser-jet printers. Some paint programs support bit-mapped (raster) or "object" (vector) modes within a single structure. It is with the text, again, that the differences are most apparent, especially with enlargement.

Clearly, there are many reasons to transform digital cartographic data between data structures. As such, we move from one type of cartographic object to another. Changing data structure can result in a loss of spatial information. These data structure transformations may not be fully invertible. Information loss can come as the result of changing scale, termed here resampling, as a direct consequence of changing data structure, or as a consequence of the data structure transformational process or algorithm itself. The study of data structure transformations, and of course their perfection as a consequence of their study, is an important goal for analytical cartography. In this chapter we will first consider scale or resampling transformations, then the specific transformation between vector and raster data structures. A classification of cartographic data structure transformations and their inverses will then be considered, and to conclude, we will discuss the role of error in data structure transformation.

10.2 RESAMPLING TRANSFORMATIONS

Resampling transformations are those in which the scale or equivalent scale of cartographic objects is changed. Resampling is usually undertaken for one of two reasons: first, so that a set of cartographic objects can be symbolized or used analytically at a scale different (usually smaller) from that at which geocoding took place; and second, to generalize a map either for clarity of symbolization or for the reduction of the data set size.

At several stages in the discussion of cartographic data structures it has been necessary to consider separately point, line, area, and volume data. Resampling transformations are within their own dimensional type, such as point-to-point, rather than between types, such as area to point.

10.2.1 Point to Point

A point-to-point transformation involves selecting one point to represent multiple points. One such point is the centroid, discussed in Chapter 9. The centroid, perhaps with a scatter parameter, is a summary of the locational characteristics of a point distribution. A primary use for the centroid is in the conversion of irregularly spaced data, or data collected throughout a set of regions, into a continuous representation such as a grid. Population density data, for example, are often computed for census tracts in a city, and converted to a grid by interpolation from point centroids selected in some manner. The point-to-grid transformation will be covered in detail in Chapter 11.

Many grid-to-grid transformations are in fact point to point, as are changes between map projections and coordinate systems. A transformational process which uses as input a global data set of latitudes, longitudes, and elevations, usually organized at regular intervals of degrees, minutes, or seconds, does not produce a regular grid after a map projection transformation. After the transformation into a map projection, the data must be resampled into a grid based on the axes of the new coordinate system. Since point-to-point transformations are so common, they are usually thought of as part of the geocoding process. They are, however, important examples of cartographic transformations.

10.2.2 Line to Line

Line-to-line transformations have received considerable attention in the cartographic literature. A general statement of the problem would be to take a cartographic line, as represented as an object in a particular data structure, and to reduce the total number of elements required to store the line in such a way that the line symbolization produced carries the spatial properties of the line to the map reader. Line character is important to preserve, yet difficult to define, and involves a complex set of related approaches to generalization (Buttenfield, 1985). The simplest techniques for reducing the number of points necessary to represent a line are nth-point elimination and equidistant resampling. In the first case, the nodes at the end of a vector line representation in entity-by-entity mode are retained as pivots, while for the remainder of the line only every nth point is kept. Similarly, for equidistant resampling, the line is followed by a distance tracking algorithm, and a point is retained at multiples of some key distance along the line. Figure 10.01 shows the difference between these methods, and the following code segment performs the generalization on a string in the STRING C language data structure introduced in Chapter 5. The function read_a_string was introduced in Chapter 5. These functions are used with the main program in Function 10.01.

Function 10.01

```
/* ========================================
/* Main program for nth-point line generalization
/* ========================================
*/
#include "header.h"
main() {
   FILE *infile, *outfile;
   int read_a_string(); void select_the_points(), write_a_string();
   /* Open an input file and an output file */
   if ((infile = fopen(INPUTFILE, "r")) == NULL)
      { printf("Unable to open file \n"); exit(); }
   outfile = fopen(OUTPUTFILE, "w");
   /* Read and generalize the strings */
   while (read_a_string(infile)) { select_the_points(SKIP);
      write_a_string(outfile); }
}
```

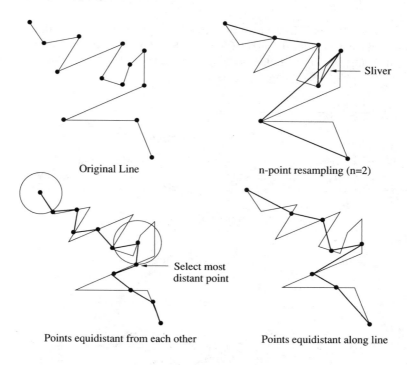

Original Line

n-point resampling (n=2)

Sliver

Select most
distant point

Points equidistant from each other

Points equidistant along line

Figure 10.01   Simple Line-to-Line Resampling Transformations

This main function assumes that the following header file has been included:

Function 10.02

```
/*  ================================================
/*    Master header file for nth-point elimination
/*  ================================================
*/
#include <stdio.h>
#include <math.h>
#define MAXPTS   5000
#define INPUTFILE "strings_input"
#define OUTPUTFILE "generalized_strings"
#define SKIP 5
struct POINT { float x; float y; };
struct STRING { int number_of_points; struct POINT  point[MAXPTS];
     } string;
```

The actual generalization is performed by the following function, select_the_points(), which selects from the input file every SKIPth point. Note that the first and last points in each string are included in the output string.

Function 10.03

```
/* ========================================================
/* Perform nth-point elimination on a string of type STRING
/* ======================================================== */
#include "header.h"
void select_the_points(skip)
     int               skip;
{
     int               j, current_point = 0;
         struct POINT     last_point;
     /* Save the last point */
     last_point.x = string.point[string.number_of_points - 1].x;
     last_point.y = string.point[string.number_of_points - 1].y;
     /* Take selected points along the string */
     for (j = 0; j < string.number_of_points; j += skip) {
         string.point[current_point].x = string.point[j].x;
         string.point[current_point++].y = string.point[j].y; }
     /* Replace the last point */
     string.point[current_point].x = last_point.x;
     string.point[current_point].y = last_point.y;
     string.number_of_points = current_point + 1;
     return;
}
```

The string can be recorded so that it can be reread, perhaps by a plotting program, by the following function.

Function 10.04

```
/* ================================================================
/* Store in an open file a string as a STRING structure
/* Note : Max x and y values is 99999.99, max # points is 99999
/* ================================================================ */
#include "header.h"
void write_a_string(outfile)
    FILE            *outfile;
{
    int             j;
    /* Write the number of points first */
    fprintf(outfile, "%6d", string.number_of_points);
    for (j = 0; j < string.number_of_points; j++)
        fprintf(outfile, "%8.2f%8.2f", string.point[j].x,
            string.point[j].y);
    /* String complete, write an end of line character */
    fprintf(outfile, "\n");
    return;
}
```

A number of techniques treat a line as well represented by the original data points selected for its representation as a cartographic object, and fit a smooth curve through the points to make symbolization more attractive or easier to interpret. Splines, polynomials, and Bezier curves are mathematical functions which have been used to perform smoothing. Buttenfield (1985) provided a list of references to the algorithms behind these functions. Many automated contouring packages use these methods to smooth the lines fed though a grid during contouring.

Among the various line generalization methods, one of the most long-standing in terms of use is the Douglas-Peucker method (Douglas and Peucker, 1973). This method uses the worst-case generalization of a line as the starting approximation, i.e., the line segment formed by simply connecting the end points. Each point along the line has an orthogonal distance from the line, which can be computed by simple trigonometry. The Douglas-Peucker algorithm selects the point with the largest orthogonal distance, breaks the line at this point, and then recursively applies this criterion to the resulting segments. Recursion continues until either only a minimum number of points remain in the string segment, or a tolerance level is reached, perhaps a proportion of the initial orthogonal distance (Figure 10.02). A number of researchers have suggested and used different criteria for evaluating line generalization methods. McMaster (1986) used a measure based upon the areas of the triangles formed between triplets of points in a line and in its generalization. Muller (1986) has suggested as a criterion the ability of an algorithm to preserve the fractal dimension of the line. The fractal dimension, a value reflecting the degree of scale invariance of a line, seems related to line complexity.

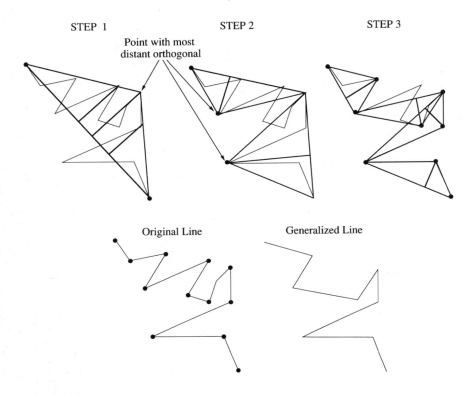

STEP 1   STEP 2   STEP 3

Point with most
distant orthogonal

Original Line   Generalized Line

Figure 10.02   The Douglas-Peucker Line Generalization Method

While line generalization by resampling reduces the number of points in the cartographic object representing the line or in its symbolization, several authors have devised
methods to actually increase the number of points. Dutton (1981) proposed an algorithm which computes midpoints for segments, and then moves the midpoints using the
values of four controlling parameters. Dutton pointed out that fractalization allows lines
to have features exaggerated, and allows the introduction of smaller scale features with
enlargement. The introduced features, being self-similar, have similar properties to the
existing line.

10.2.3 Area to Area

Resampling areas to yield areas is a resampling transformation of considerable value to
analytical and computer cartography and to GIS. A general statement of the goal is to
merge multiple data sets into a set of regions such that the merged data allow
comparison between maps. Usually, at least for sets of geographic areas, this means
computing a set of greatest common geographic units, or areas which need no longer be
partitioned to represent spatial data as cartographic objects. As a practical expression,

consider non-nested regions such as census tracts and police districts. A crime study may wish to collate population characteristics by police district, only to find that the boundaries do not coincide with census tracts. Similarly, when data have to be compared between different time periods, invariably change in the geographic extent of regions has taken place.

How analytical and computer cartography deals with this problem is largely a function of the data structure used to store the various map layers, and the data structure in which the map comparison is to be conducted. Clearly, for two maps in two different structures there are two strategies. First, we can transform both maps into a common data structure which allows comparison directly. Often this is done by converting to a set of topological or entity-by-entity most common geographic units, or to a grid. In the case of the grid, really we perform an area-to-point transformation, so that the points coincide for two or more sets of regions. The second strategy is to convert one of the maps into the structure of the other. Thus a polygon entity-by-entity map can be gridded to compare with another grid. As the number of layers increases, the first of these strategies becomes more suitable, especially when inverse transformations are required.

Overlaying two maps to generate a set of most common geographic units in vector mode is not a trivial task (Goodchild, 1978). Central to the problem is the processing of line data to determine intersection points between overlapping polygons, so that they can be added to both polygons in the correct place. The points then become nodes in the network of lines which constitute the least common geographic units. A summary of the contributions of computational geometry to these problems is the book by Preparata and Shamos (1985). By the late 1970s, polygon overlay was available within a number of GISs and automated mapping systems (Franklin and Wu, 1987), especially within the Whirlpool module of Odyssey (White, 1978).

The procedure for polygon overlay, illustrated in Figure 10.03, consists of three separate sub-problems. First, the intersection points between lines must be found. Usually, this is done using a line segment intersection routine such as that shown in Chapter 9 recursively for all pairs of line segments in the two chains. A way to save considerable computation time is first to check the bounding rectangles of the two chains for overlap. If there is no overlap between bounding rectangles, then there is no need to test each line segment for intersection. When intersections are found, the chains must be split, and the intersection point must be labeled as a node and included as the end point of the new sub-divided chains. Many mapping systems compute and save the map coordinates of the bounding rectangles of lines and areas automatically on data entry for this purpose.

In the next stage, the partitioned chains are reassembled into the new set of polygons which make up the most common geographic units. Finally, each polygon must be relabeled; either with a new set of sequential or other labels or with labels which record the partitioning history. Particularly difficult to process are polygons which cross and recross boundaries, and islands. A FORTRAN implementation of polygon overlay was published by Baxter (1976). Numerous improvements and refinements have been reported over the last few years, and the polygon overlay problem remains the topic of considerable work in analytical cartography.

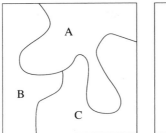 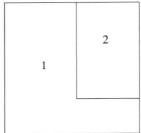 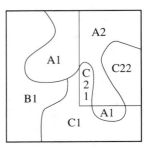

Figure 10.03   The Polygon Overlay Problem

10.2.4 Volume to Volume

Volumetric representation is rare in cartography, since there are few truly three-dimensional means by which to symbolize cartographic objects. The usual volume-to-volume resampling transformations are changes in the grid spacing associated with a grid data structures, or changes in a TIN surface representation. Within a TIN, rarely is the set of points originally used to depict the surface changed, since the points are locations of real data observations, and the data structure is fairly compact in the first place. When TINs are compared to data in other formats, such as grids, either the TIN is interpolated to a grid, or the data are points and are allocated to TIN triangles using a point-in-polygon test.

Grids are, however, frequently resampled to change size, to retrieve a subset, or to change the grid spacing. Invariably, unless the change is simple, such as taking every other row and column, the new grid is generated by computing the location of the new grid intersections in the coordinate space of the map, and then by interpolation from neighboring grid cells. Usually, the four neighbors and a simple unweighted average are sufficient for the resampling.

10.3 VECTOR TO RASTER AND BACK AGAIN

We have seen in the previous chapters that data structures for analytical and computer cartography often reflect the demands of the hardware and software they support. Just as display devices can be categorized as vector or raster, so can the cartographic data structures used by software. In the past, a broad division was made between raster and vector data structures. Each structure has its advantages and disadvantages, and has also had its proponents. The vector-versus-raster debate, however, has become less of an issue since the development of algorithms for efficient data structure transformation. The reasons for transformation are many, and in many automated mapping systems the conversion of raster to vector data, or vice versa, is a major consumer of computer time. The raster structure is a grid format, and stores a map as individual pixels on lines similar to a television picture and a satellite image. The vector structure stores maps line by line, feature by feature. The vector format is most suitable for storing as objects and symbolizing the cartographic entities familiar to human thought (Peuquet, 1979). Most cartographic objects represent boundaries, rivers, coastlines, railroads, etc.; i.e., distinct lines or groups of lines. On the other hand, raster devices operate reliably, flexibly, and have advanced data handling capability. For data storage, the manipulation of data in raster mode is quite straightforward.

The repertoire of vector algorithms for cartographic transformations was more quickly developed than for raster algorithms (Peuquet, 1979). This difference, however, rapidly diminished during the 1980s. Normally, spatial data stored in vector formats take less storage space than does their raster mode equivalent. The graphic output devices in vector form are relatively accurate and distinct, but the speed of output depends on the length of lines on a map.

The greatest need for transformation comes from the fact that automatic scanners produce raster data, while vector digitizing requires human control. Therefore, to systematically capture large data sets for cartography, the conversion from a raster data structure to a vector form becomes necessary.

10.3.1 Vector to Raster

The conversion of vector data to raster or grid form is usually termed *rasterization*. Rasterization involves four distinct steps. First, the vector data must be read into the computer for processing. The processing can take place one polygon or line at a time, or simultaneously for a whole map. Second, the appropriate scales and map transformations should be applied. Normally, the map projection transformation and any resampling transformation will have taken place before this step is reached; for example, a world map may have been transformed to a Mercator projection, clipped at 84 degrees north and 80 degrees south, and the map resampled using the Douglas-Peucker method. We can now assume that the alignment of the grid will be to the axes of the coordinates used by the input data. The only remaining decision is how many pixels the map will be converted into, and this is determined either by the desired map resolution or the number of rows and columns. Rectangular pixels are sometimes desired, especially to meet the demands of a particular display device, algorithm, or application. This step establishes the geometry of the rasterization. The third stage depends upon technique. In many cases, the grid is partitioned as an array in the computer's memory, and as the points, lines, and polygons are rasterized, the 0s stored initially are changed to 1s, or to an index number for the line or polygon. For very large arrays a second method is used. In this method, the locations of non-zero pixels, with their indices if necessary, are stored, either as a file or internally in [row, column, index] format. These data are then sorted by row and column, and an array is constructed by reading the sorted data, filling rows and columns either with the stored value or with a zero. An option is to fill a polygon with an index value. The final step is to store the array in the required format, such as with run-length encoding or as a bit map.

A number of steps can be performed to speed the rasterization process. First, if the distance between two points in the original coordinate system is smaller than half the grid spacing, the second point can be eliminated. This reduces unnecessary data and saves processing time. A careful choice of sorting technique can also make a large difference in the processing time used. Another step is to process the vector data, computing and storing linear equations of each two points. The purpose of the step is to compute in advance the linear equation constants for each line segment. These can be stored as real numbers, though Bresenham (1965) has shown that only integers are necessary. One problem is the special case of vertical lines, which have an infinite gradient and tend to be common on maps. In this case, the programmer can store the fact that the line is vertical.

The rasterization is now complete, yet for symbolization, the line often needs to be thickened. Simple thickening can parse the whole array and change from zero to one (no line to line) any pixel which borders a line pixel. Unfortunately, this tends to fill in the fine details of wiggly lines. Another technique, which in some display devices can be performed in hardware, is to fill in these neighboring pixels with values to be displayed at a lower light intensity than the line pixels themselves, a technique known as anti-aliasing. Anti-aliasing removes some of the effect of stair-stepping along diagonal lines in raster mode, an effect usually called the "jaggies".

The following functions can be used together to convert from vector to raster data.

Function 10.05

```
/* ==============================================
/* Main program for v2r: vector to raster
/* ============================================== */
#include "header.h"
main() {
     void scale(), vectoras(), record();
     scale();     vectoras();     record();
}
```

Function 10.05 is a main program which simply calls a sequence of other functions. The master header file is given as Function 10.06, and uses the POINT, STRING, and GRID structures defined in Chapter 5.

Function 10.06

```
/* ================================================
/*    Master header file for v2r
/* ================================================ */
#include <stdio.h>
#include <math.h>
#define MAXPTS   500   /* The maximum number of points in a string */
#define MAXROW   100   /* The maximum number of rows in the grid */
#define MAXCOL   100   /* The maximum number of columns in the grid */
#define MAXLABEL 10    /* Maximum number of characters in the grid
                          descriptor (Unused here) */
#define INPUTFILE "yourinputfile" /* Path and filename for your
                                     input file */
/* Zero dimensional objects */
struct POINT { float x; float y; };
/* One dimensional objects - Geometry Only */
struct STRING { int number_of_points; struct POINT    point[MAXPTS];
     } string;
struct GRID {
     char            grid_descriptor[MAXLABEL];
     int nrows; int ncols; struct POINT    corners[4]; int datum;
     int             z[MAXROW][MAXCOL]; } grid;
float            xscale, yscale;
float            segment[MAXPTS - 1][4];
```

In the first of the major functions, the user is prompted for the world coordinate limits
of the area on interest, and the number of rows and columns required.

Function 10.07

```
/* ============================================== */
/* Function to prompt for coordinate information */
/* ============================================== */
#include "header.h"
void scale() {
     /* Prompt for necessary coordinate & pixel information */
     fprintf(stdout, "\n Enter [x,y] at lower left of vector
         map :");
     fscanf(stdin, "%f%f%*1c", &grid.corners[0].x,
         &grid.corners[0].y);
     fprintf(stdout, "\n Enter [x,y] at upper right
```

```
        of vector map:");
    fscanf(stdin, "%f%f%*1c", &grid.corners[2].x,
        &grid.corners[2].y);
    fprintf(stdout, "\n \n Grid lower left will be zero
        indexed [0,0]\n");
    fprintf(stdout, "\n Enter number of columns and
        rows [col,row] :");
    fscanf(stdin, "%d%d%*1c", &grid.ncols, &grid.nrows);
    xscale = (grid.corners[2].x - grid.corners[0].x) /
        (float) grid.ncols;
    yscale = (grid.corners[2].y - grid.corners[0].y) /
        (float) grid.nrows;
    printf("\nScaled area will be %4d pixels wide by
        %4d pixels high\n", grid.ncols, grid.nrows);
    printf("One grid cells equals %f units in x and
        %f units in y\n", xscale, yscale);
    return;
}
```

Vector data are read and inserted into the appropriate places by using the function store_a_string(), which computes the linear equation constants for the line and provides them through the variable segment[][] to the remainder of the program. The function store_a_string() is referenced by the function vectoras(), listed below.

Function 10.08

```
/* =============================================== */
/* Read and derive linear equations for one string */
/* =============================================== */
#include "header.h"
int store_a_string(infile)
    FILE    *infile;
{
    int     j, vertical_line, xdif, ydif; float alpha, beta;
    if (fscanf(infile, "%d", &string.number_of_points) != EOF) {
        for (j = 0; j < string.number_of_points; j++)
            fscanf(infile, "%f%f", &string.point[j].x,
                &string.point[j].y);
        /* Calculate Linear Equations for Segments    */
        for (j = 0; j < (string.number_of_points - 1); j++) {
            alpha = beta = 0.0; vertical_line = 0;
            if ((xdif = (int) (string.point[j].x -
                string.point[j + 1].x)) != 0) {
                beta = (string.point[j].y -
```

```
                                string.point[j + 1].y) / (float) (xdif);
                    alpha = string.point[j].y -
                            beta * string.point[j].x;
            } else vertical_line = 1;
            segment[j][0] = vertical_line;
            segment[j][1] = alpha; segment[j][2] = beta;
            ydif = (int) (string.point[j].y -
                    string.point[j + 1].y);
            segment[j][3] = (float) sqrt((double)
                    (xdif * xdif + ydif * ydif));
            /* Stored values in array segment:
            /* [0] vertical_line 0=ok 1=segment is vertical
            /* [1] alpha = intercept  y = a + b x
            /* [2] beta = gradient    y = a + b x
            /* [3] distance = length of segment  */
        } return (1); } else return (0);

}
```

The vector data are next converted to a grid using the information provided. Small increments of distance are used so that every grid cell along the line will be included, simply by virtue of it touching the line. The sampling interval is chosen so that every cell will be included. This is a different approach from the more sophisticated technique of determining if each new grid cell along the line is necessary to preserve connectivity. The result is "fat lines" when lines are wiggly or diagonal but close to vertical or horizontal (Figure 10.04). Assignment of a 1 to a grid cell in this case is by dominance, i.e., the cell contains a significant length of the line, with the constraint that in some cases non-dominant cells must be included to avoid breaking the line.

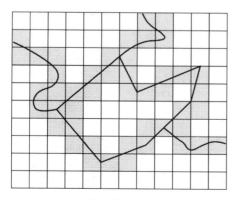
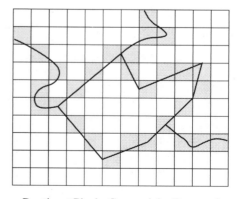

Any Pixel Intersected   Dominant Pixel - Connectivity Preserved

Figure 10.04   Two Approaches to Cell Filling in Vector-to-Raster Transformation

This can lead to topological blunders, such as holes (Figure 10.04) where none exist, unlike the simpler method, which simply fills the whole area around an intersection. This effect could be eliminated by line thinning, discussed below.

The function which rasterizes the data is vectoras(), which in turn uses a function store_a_string() above. Vectoras fills a grid structure with 1 where a line is detected and with 0 elsewhere. A refinement would be to include the option to label the grid with the number of the string, an attribute of a string, or to fill the area enclosed by a RING with index numbers or attributes for a polygon.

Function 10.09

```
/* ===================================== */
/* Function to Rasterize String Data */
/* ===================================== */
#include "header.h"
void vectoras()
{
    FILE    *infile; int i, j, string_no = 0, xindex, yindex;
    float   xbegin, ybegin, xend, yend, temp;
    float   step, offset, x, y, mindist, xbegin, ybegin, xend,
        yend, temp;
    /* Initialize array raster to zeros        */
    for (i = 0; i < grid.nrows; i++) for (j = 0; j < grid.ncols;
        j++) grid.z[i][j] = 0;
    /* Determine minimum distance at which to generate pixels  */
    /* so that no gaps occur (according to sampling theorem)   */
    mindist = xscale; if (yscale > mindist) mindist = yscale;
        mindist /= 2.0;
    /* Begin process, working one polygon at a time  */
    if ((infile = fopen(INPUTFILE, "r")) == NULL) {
        printf("Unable to open file \n"); exit(); }
    while (store_a_string(infile)) {
        printf("Processing string %d with %d points\n",
            ++string_no,string.number_of_points);
        /* Compute pixel locations along each segment    */
        for (j = 0; j < string.number_of_points - 1; j++) {
            xbegin = string.point[j].x;
                ybegin = string.point[j].y;
            xend = string.point[j + 1].x;
                yend = string.point[j + 1].y;
            /* If the segment is reversed, switch nodes     */
            if (xend < xbegin) {
                temp = xbegin; xbegin = xend; xend = temp;
                temp = ybegin; ybegin = yend; yend = temp; }
            for (step = 0.0; step <= segment[j][3];
                step += mindist) {
```

```
                     /* Deal with the vertical segments    */
                     if ((int) segment[j][0]) {
                         if (ybegin > yend) {
                             temp = ybegin; ybegin = yend;
                                 yend = temp; }
                         x = (float) xbegin;
                                     y = (float) ybegin + step;
                     } else {
                         /* Calculate projection onto x axis
                             of the step */
                         offset = step * ((float) (xend - xbegin)) /
                             segment[j][3];
                         y = segment[j][1] + segment[j][2] *
                             ((float) xbegin + offset);
                         x = xbegin + offset; }
                 xindex = (int) ((x - grid.corners[0].x) /
                     xscale);
                 yindex = (int) ((y - grid.corners[0].y) /
                     yscale);
                     if ((xindex >= 0) && (xindex < grid.ncols)
                         && (yindex >= 0)
                         && (yindex < grid.nrows))
                         grid.z[yindex][xindex] = 1;
             }
         }
     }
     printf(" Processing Complete \n");
     return;
 }
```

Finally, the function record() writes the array to a disk file for display or further analysis. This function is the opposite of read_a_grid(), listed in Chapter 5.

The program resulting from linking and compiling these modules was used with a world data base and the mercator() function from chapter 9 to rasterize part of North America (Figure 10.05).

10.3.2 Raster to Vector

Conversion of vector data to raster is comparatively simple compared to the inverse transformation. However, the demand for the raster-to-vector transformation is increasing as raster input devices such as scanners and remote sensing devices become more widespread. Increasingly, therefore, the raster-to-vector transformation is being built into automated mapping systems and GISs. This process is very CPU intensive, and can yield topological and other errors in the resultant vector data set regardless of the

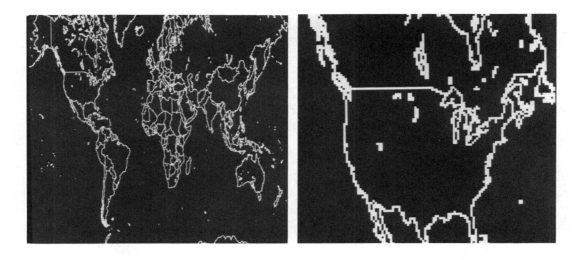

Figure 10.05   Rasterized Vector Map of North America

quality of the input data. In general, analytical cartography has been slow to research issues related to this important transformation, and as a result the published work is mostly in image processing. Peuquet (1981) noted the lack of efficient algorithms and poor computer coding in this area, a deficiency which has not been fully addressed.

Converting from raster to vector data involves four steps. These stages, shown graphically in figure 10.06, can be termed *skeletonization* or *line thinning*, *line extraction*, and *topological reconstruction*. Line thinning is necessary because vector lines are purely geometric, i.e., they have zero width as cartographic objects, yet have a width determined by the grid cell size when they appear in raster mode. The grid representation of the lines, therefore, must first be thinned so that the line consists merely of a one-cell-width sequence of connected pixels. This connectivity is usually in one of the eight major directions dictated by the eight-cell connectivity of grid cells.

Line extraction, the second stage, involves determining where individual lines begin and end in the thinned image, so that the lines can be rewritten as vectors connecting the end points in the correct sequence. Points are usually generated at the centers of the line grid cells, and long straight sequences can be eliminated to reduce the number of points in the line. Topological reconstruction is the process of generating the topological connectivity of the lines to rebuild a topological definition of the lines and polygons from the entity-by-entity objects which come from the previous two stages. As such, the third step is identical to the transformations from entity-by-entity to topological data structure discussed in the next section.

The first stage, skeletonization, can be performed in one of three ways. Peuquet termed these peeling, expanding, and medial axis. The medial axis method is the fastest, but it works on one line at a time and not on the whole grid simultaneously, and inconsistencies occur in very thick lines. Peeling and expanding methods are the

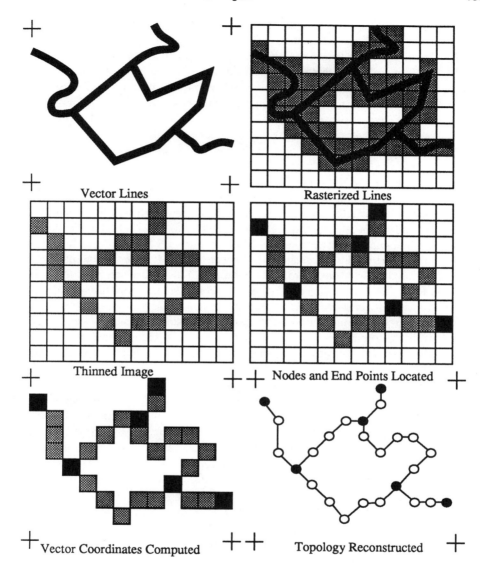

Vector Lines

Rasterized Lines

Thinned Image

Nodes and End Points Located

Vector Coordinates Computed

Topology Reconstructed

Figure 10.06   Stages in Raster-to-Vector Conversion

inverse of each other. Peeling deals with systematically eroding the edges of lines, while expanding deals with expanding the space between lines instead.

A highly efficient peeling method is Pavlidis's asynchronous thinning algorithm (Pavlidis, 1982). However, this technique is not completely parallel in operation and in order to maintain connectivity accepts lines with a width of more than one grid cell. Rosenfeld and Kak's thinning algorithm (Rosenfeld and Kak, 1981) is as fast as asynchronous thinning and results in a line of single-grid-cell width. This method works iteratively, at each pass deleting border grid cells which do not disconnect the local (3 by 3 cell) neighborhood. The result of this first step is a single-cell-width line, but sometimes the thinning process introduces artifacts into the geometry of the lines.

Line extraction can be accomplished in one of two ways. Line following seeks a node which forms the beginning of a line, and attempts to follow the line to another node. One of its disadvantages is that when a node is reached and the line terminated, no use is made of the fact that other lines normally begin at this point. The scan-line approach uses the same logic as line following, but processes multiple lines simultaneously.

Once the lines have been extracted, they are usually written as vector lines and any topological processing takes place in vector mode. The write to vectors involves finding the absolute or world coordinate location of each grid cell, which can be accomplished by taking the grid cell row or column number, dividing by the number of rows or columns respectively, multiplying by the size of a grid cell in that direction, and adding the world coordinate of the lower left-hand corner of the map. When the grid is not aligned with the coordinate system, the four values stored in the grid.corners[] part of the GRID structure can be used to compute the affine transformation necessary to convert to world coordinates.

When point data are processed rather than vectors, the coordinate transformation is the only one which need be applied. A special case is when the data to be processed are coverage polygons without boundaries. Data such as spectral classification and clustering of remote-sensing data or data from existing raster mode GISs are often in this format. An approach to finding the boundaries is to scan the grid from top to bottom and from side to side to reveal the boundaries. A simple algorithm to do this is illustrated in Figure 10.07. Alternatively, the grid cell can be filtered using an edge detecting filter, which emphasizes breaks in value. A disadvantage of this method is that one row and one column from each edge of the image are lost for every time the filter is applied. Clearly, when a grid cell is wholly within a polygon, the grid cell is assigned the index for that polygon.

This process is the exact inverse of the way in which polygons are created as grids from vector data. In the vector-to-raster section above, we considered only line rasterization. Points can be gridded simply as individual grid cells, but areas are different. When cells fall into two or more polygons, one way of making the assignment is to compute the area of the cell occupied by each polygon, and to assign the cell to the polygon which occupies the most area. Alternatively, the polygon boundaries can be processed as above, and the indices for the polygon interior assigned by filling the bounded area within each polygon. In this case, the boundaries would remain as such, and would not be considered part of the polygon. This distinction is made in the digital cartographic data standards, between the ring defining a polygon and the polygon's interior area. For applications where input data are classified remotely sensed imagery, the lack of boundaries is normal, as also is a profusion of small and even single-cell clusters, which would make poor vector equivalents. These can be eliminated by filtering,

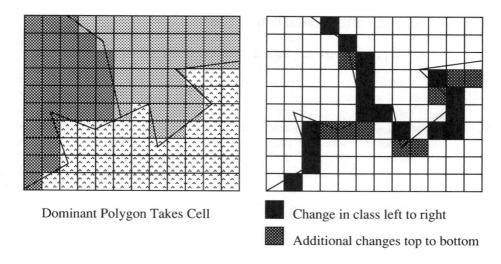

Dominant Polygon Takes Cell   Change in class left to right

Additional changes top to bottom

Figure 10.07   Detection of Edges in Classified Images

or by assigning small clusters to larger, neighboring clusters if certain criteria are met, such as diagonal connectivity. As remotely sensed data get better resolution, the connecting problem will diminish in significance, although it will remain for small-scale mapping, such as land use coverage. However, the elimination of finer detail or "salt and pepper" will become more important with higher resolution (Figure 10.08).

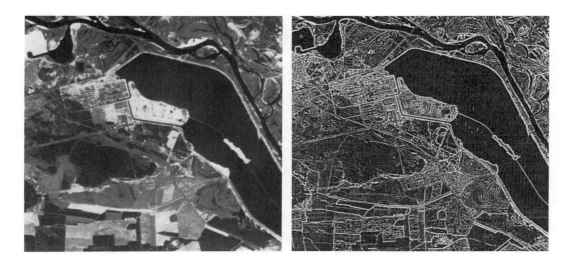

Figure 10.08   Edge-Filtered Image Showing Resolution Problems

10.4 DATA STRUCTURE TRANSFORMATIONS

10.4.1 A Framework for Data Structure Transformations

Data structure transformations are usually necessary because of the demands of a particular mapping system, and as such are pertinent to computer cartography. However, the need for efficient transformations, and the in-depth understanding of the cartographic implications of the transformations, are very much a part of analytical cartography. The multitude of data structure transformations performed during the mapping process fall into one of three types. Of these, type A are transformations of cartographic objects by their scale alone. Data structure transformations such as line generalization and grid resampling fall into this category. Resampling transformations, such as the rewriting of cartographic data in the same data structure at a different resolution, are the digital expression of the scale phenomenon. The study of such scale transformations has made important contributions to analytical cartography.

A second type of cartographic data structure transformation (Type B) involves a dimensional change. A dimension change is a change in the type of cartographic object itself rather than a data structure transformation. Such a transformation is a logical data compression. For example, a dimensional data structure change may be the selection of a point to represent an area. Such a transformation involves considerable loss of data and information, yet gives the clarity or simplicity sometimes required during cartographic generalization. Such a transformation is invertible. We could, for example, generate Theissen polygons from the points to take the place of the original polygons, but the inverse transformation is an imperfect one and results in error. These types of transformations were considered in Chapter 9 as map data transformations, since they involve actual manipulation of the spatial properties of the cartographic objects.

Type C data structure transformations move cartographic objects between structures as part of the mapping process, without much change of scale. For example, a gridded digital elevation model could be processed for significant points such as peaks, saddles, and pits, from which to extract and generate a TIN. The areal coverage is identical; essentially the same cartographic object, the terrain, is available for symbolization or analysis, the object dimension (3) remains the same, yet the data structures are radically different.

The four major data structures we have covered can be classified into entity-by-entity, topological, TIN, and grid. The grid includes all raster-based structures such as quad trees, and the entity-by-entity covers some of the hybrid structures, such as Freeman codes. Given these four types of data structure, a matrix of transformations can be compiled which shows the 16 possible transformations of type C (Figure 10.09). The leading diagonal of this matrix involves transformations without a change of data structure, and as such these transformations must be type A or type B. Of the remaining 12 cells in the matrix, four fall into the grid-cell to entity-by-entity and topological structures and their inverses, which were considered above as raster-to-vector and vector-to-raster transformations. The remaining eight cells in the matrix are four transformations and their inverses. These are entity-by-entity to topological, entity-by-entity to TIN, topological to TIN, and TIN to grid.

	Entity by Entity	Topological	TIN	Grid
Entity by Entity	Type A Type B	Type C	Type C	Vector to Raster
Topological	Type C	Type A Type B	Type C	Vector to Raster
TIN	Type C	Type C	Type A Type B	Type C
Grid	Raster to Vector	Raster to Vector	Type C	Type A Type B

Figure 10.09   Possible Type C Data Structure Transformations

The first of these transformations, entity-by-entity to topological, is the normal way by which topology is added during the geocoding process. When maps are digitized, the topology can either be entered explicitly, as in the DIME files, as separate records, or it can be extracted from the points, lines, and polygons as they are digitized. For strings, topological attributes are forward and reverse linkages and the labels of the neighboring polygons. The linkages to other strings can be determined by storing separately the beginning and ending nodes of the strings, and then matching them against each other. Since manual digitizing results in small differences in the exact coordinate values at points, usually some tolerance is used, and matching points are given the same, average location (snapping) (Figure 10.10). Many digitizing programs maintain and check this information as each string is entered, and prompt the user if an end-point match cannot be found with the existing strings. The end-point and topological information can be stored in a structure such as the CHAIN, introduced in Chapter 5.

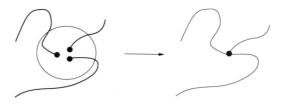

Figure 10.10   Node Snapping in Entity-by-Entity to Topological Transformations

Finding the names of the neighboring polygons is more difficult, and the usual method is either to prompt the user for the left and right polygon information when a new chain is accepted, or to have the user digitize a set of label points, one per polygon. These label points can be stored separately, for example with the attribute data for the polygons. When polygons need to be referenced, or when attributes are required in the symbolization transformation, the label points can be tested using a point-in-polygon test for inclusion within a group of chains. The relevant data structures are the CHAIN, the AREA_CHAIN, and the NETWORK_CHAIN.

The inverse of this transformation is the extraction of the simpler structures, such as STRINGs, ARCs, and RINGs, from the topological structures. Since the to_node and from_node information is already available, this conversion is simply a rewriting of the information already contained within the structure. In some cases, chains would have to be written into the RINGs backwards, i.e., from end_point to begin_point, so that the sequential nature of the RING is maintained. This is especially important if the motive for the transformation is to symbolize the polygon as a filled area under GKS, or if the polygon area is to be computed, as in Chapter 9.

The six remaining transformations all involve the TIN structure. They are topological to TIN, entity-by-entity to TIN, and TIN to grid and their inverses. To date, no real example of a topology-to-TIN transformation is available, although the inverse, in which the TIN is used to generate a stream, ridge line, or connected set of polygons representing visible or invisible areas, seems analytically valuable. This transformation is closer to class B, since the topological data structure is usually two-dimensional, while the TIN is three-dimensional. The network, having no partial triangles involved, could be generated using directed links, with direction being downhill, for streams or ridge lines. For intervisibility problems, however, the splitting of triangles into polygons would be necessary, and simple or complex polygons would have to be formed to store the visible or invisible areas.

The transformation from entity-by-entity to TIN is also of class B, since the only real example of this transformation is the transformation from point entity-by-entity to TIN, and vice versa. The forward transformation in this case is Delaunay triangulation, while the inverse is again simply a relisting of the entities contained in the TIN structure. Finally, the grid-to-TIN forward and reverse transformation is a true class C transformation, since neither dimensional change nor resampling takes place. This transformation can be accomplished in the forward case by selecting significant points, perhaps by filtering or special case searching, which with their elevations form the basis of the resultant TIN. The inverse transformation is accomplished by linear or other interpolation of elevations at grid cell locations from the TIN's triangles.

This three-class classification of the transformations possible among the four major cartographic data structures can be used to understand the interrelationship among the power, flexibility, efficiency, and storage size of data structures, and helps in determining when data structure transformations are appropriate within a mapping system. Analytical cartography can assist in this understanding by allowing the cartographer quantitatively and theoretically to model and predict the amount and distribution of error to be expected as a result of data structure conversions. The contention is that a cartographic entity is a complex phenomenon, and the means by which this entity is converted to a cartographic object as digital data within a cartographic data structure are both controllable and predictable. With correct understanding of these transformations, cartographers should be able to provide error models and reliability estimates along with maps which are the geographic equivalent of the statistician's normal curve and tests of significance.

10.5 THE ROLE OF ERROR

Much recent research in cartography has been centered on the problem of errors in digital cartographic data bases. The focus of the problem is that since digital cartographic data bases can be highly precise, many users of the maps produced from the data bases also believe them to be accurate. In fact, the data stored as digital maps are often unbelievably faithful reproductions of incorrect maps. Beard (1989) classified map errors into *source errors*, *use errors*, and *process errors*. Source errors are errors in the original map sources, in the digital cartographic source data, or in the data capture and geocoding processes. Use errors are caused by a lack of information about a mapping system, unexpected deviation from cartographic convention, the use of maps with a scale which is too small, a lack of timeliness, and a lack of user documentation. Beard proposed that increased attention be devoted to use error, especially since users of GIS are typically not always cartographers.

Part III of the digital cartographic data standards is designed to ensure that the source errors are revealed. The standards propose a "truth in labeling" approach, which also makes the provider of digital cartographic data responsible for providing, either internally or externally, documentation in the form of a quality report. The quality report should contain information on lineage of the data, on positional accuracy of the data, on the accuracy of the non-map attributes associated with the data, and on the completeness of the data. Part IV of the digital cartographic data standards assists in the task of reducing error by providing a standard set of entity and attribute terms for cartographic features which, if used, reduce ambiguities about digital cartographic data. Also, Part II of the standards establishes the nomenclature and definitions for digital cartographic data transfer, so that data acquired from government or other sources will be in a consistent structure, as documented in the standards.

The final type of error in Beard's classification is process error, the error resulting from the digital conversion, scale change, projection change, or the type of symbolization used. These errors are those attributable to one or more of the cartographic transformations described in this and the preceding two chapters. This type of transformational error clearly relates to the types of transformation discussed above. As we have seen, cartographic transformations can be map-based or data structure-based, and can involve changes in resolution and changes in object dimension. Three types of process error can therefore occur.

First, errors can occur in the geometry of the map, that is, features can become mislocated, in the three-dimensional geometry of the world. These mislocations are sometimes unknown, but can be the result of a controlled distortion, such as changing the map projection or statistical space fitting. Second, the errors can be due to scale change. An ideal situation would be to have a single, very high resolution data base from which all others are generated using some objective generalization function (Beard, 1987). This is rarely the case, and maps will continue to be digitized from a large number of sources at different scales. Again, some errors are unknown, such as the removal of islands as they become smaller and smaller with scale, but others, such as the gridding error associated with vector-to-raster conversion, are measurable. In many cases, cartographers have suggested means for defining and measuring errors in these cartographic transformations so that they can be reduced (McMaster, 1986).

Third, errors occur due to the transformation of cartographic objects in digital form between data structures. The case can be made that none of these errors are due to unknowns, since the conversion is under the full control of the cartographer. With the encoding of topology, many consistency checks can become integral parts of GISs and computer mapping systems, ensuring that cartographic errors are more simply detected and corrected (Wagner, 1988). This approach is an integral part of the TIGER files discussed in Chapter 4.

Errors due to changes in data dimension are substantial, but are deliberate in the sense that they allow analysis or display which would otherwise be impossible. It is the straight data structure transformation errors which are most obviously manageable, and part of analytical cartography is clearly the pursuit of data structure transformations which minimize, or at least give accounts of, the error they introduce. The digital cartographic data standards suggest the use of a quality map, one which maps out the error expected in the cartographic data. It is only by fully understanding and modeling cartographic error that cartographers can ensure that their products survive the tests of time as today's digital cartographic data bases become the historical archives of the future.

10.6 REFERENCES

BAXTER, R. S. (1976) *Computers and Statistical Techniques for Planners*, Methuen, London.

BEARD, M. K. (1987) "How to survive on a single detailed database", *Proceedings, AUTO-CARTO 8*, Eighth International Symposium on Computer-Assisted Cartography, Baltimore, MD, March 29-April 3, pp. 211-220.

BEARD, M. K. (1989) "Use error: the neglected error component", *Proceedings, AUTO-CARTO 9*, Ninth International Symposium on Computer-Assisted Cartography, Baltimore, MD, April 2-7, pp. 808-817.

BRESENHAM, J. E. (1965) "Algorithm for computer control of a digital plotter", *IBM Systems Journal*, vol. 4, no. 1, pp. 25-30.

BUTTENFIELD, B. (1985) "Treatment of the cartographic line", *Cartographica*, vol. 22, no. 2, pp. 1-26.

DOUGLAS, D. H. AND PEUCKER, T. K. (1973) "Algorithms for the reduction of the number of points required to represent a digitized line or its caricature", *Canadian Cartographer*, vol. 10, pp. 110-122.

DUTTON, G. H. (1981) "Fractal enhancement of cartographic line detail", *The American Cartographer*, vol. 8, no. 1, pp. 23-40.

FRANKLIN, W. R. AND WU, P. Y. F (1987) "A polygon overlay system in PROLOG", *Proceedings, AUTOCARTO 8*, Eighth International Symposium on Computer-Assisted Cartography, Baltimore, MD, March 29-April 3, pp. 97-106.

GOODCHILD, M. F. (1978) "Statistical aspects of the polygon overlay problem", *An Advanced*

Study Symposium on Topological Data Structures and Geographic Information Systems, Laboratory for Computer Graphics and Spatial Analysis, Harvard University, Cambridge, MA.

McMaster, R. B. (1986) "A statistical analysis of mathematical measures for linear simplification", *The American Cartographer*, vol. 13, no. 2, pp. 103-116.

Muller, J.-C. (1986) "Fractal dimension and inconsistencies in cartographic line representations", *The Cartographic Journal*, vol. 23, pp. 123-130.

Pavlidis, T. (1982) "An asynchronous thinning algorithm", *Computer Graphics and Image Processing*, vol. 20, pp. 133-157.

Peuquet, D. J. (1979) "Raster processing: An alternative approach to automated cartographic data handling", *The American Cartographer*, vol. 6, no. 2, pp. 129-139.

Peuquet, D. J. (1981) "An examination of techniques for reformatting digital cartographic data. Part 1: The raster-to-vector process", *Cartographica*, vol. 18, no. 1, pp. 34-48.

Preparata, F. P. and Shamos, M. I. (1985), *Computational Geometry*, Springer-Verlag, New York.

Rosenfeld, A. and Kak, A. C. (1981) *Digital Picture Processing*, Academic Press, New York, Second Edition.

Wagner, D. F. (1988) "A method of evaluating polygon overlay algorithms", *Technical Papers*, ACSM-ASPRS Annual Convention, vol. 5, pp. 173-183.

White, D. (1978) "A new method of polygon overlay", *Harvard Papers on Geographic Information Systems*, vol. 6, Harvard University, Cambridge, MA.

11

Terrain Analysis

11.1 THREE-DIMENSIONAL TRANSFORMATIONS

The cartographic transformations discussed so far have avoided the instances where three-dimensional data are involved. In analytical and computer cartography, a large quantity of digital data concerns the third dimension, usually the surface of the land or the bottom of bodies of water. These data are three-dimensional since we need both eastings and northings and the elevation at a location to give the necessary level of information. The part of analytical cartography concerned with the analysis of terrain type data, and here we can include any continuous surface, is called terrain analysis. Here, we have broadened terrain analysis to include the symbolization of terrain using computer cartography. Both analytical and computer cartography must be concerned with the data structures and the special purpose data structure conversions involved in terrain analysis.

The first of these data structure conversions is by far the most critical and can fully determine the level of accuracy and the error involved in terrain analysis and mapping. This transformation is that of point data to both the TIN and the grid three-dimensional formats. In only a few instances are terrain data collected at locations which are evenly spaced. Usually, elevations are measured at significant points on the landscape, such as the crests of ridges, the tops of hills and mountains, and the bottoms of depressions and lakes. When data are collected in a grid, such as the elevations taken from stereophoto interpretation using an analytical stereoplotter, there is often a need to treat the data as point data for transformation into a particular map projection or spacing, and then to reinterpolate to a grid. The point-to-TIN and point-to-grid conversion is clearly, therefore, important. In a TIN, the irregularly spaced points can be used directly as part of the three-dimensional cartographic object definition, eliminating the interpolation problem. For the grid, however, there is no one simple solution to the interpolation problem, and we often are left making trade-offs to achieve terrain which suits a cartographic purpose.

11.2 INTERPOLATION TO A GRID

A general statement of the interpolation problem would be: Given a set of point elevations with coordinates (x, y, z), generate a new set of points at the nodes of a regular grid so that the interpolated surface is a reasonable representation of the surface sampled by the points. Figure 11.01 shows a map containing such a set of points. The data for these points are presented below in Function 11.01.

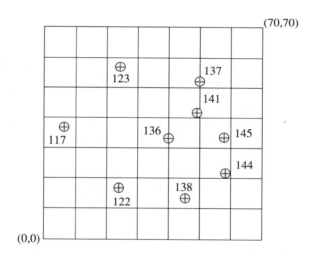

Figure 11.01   Test Point Data Set

Units for x and y are often meters, but can also be degrees, feet, etc. Units for the z value are usually meters, rounded to the nearest meter, or feet. Ocean depths are often meters, fathoms, feet, etc. In the following worked examples, the test data set above will be converted to a grid. The grid starts at (0,0) and ends at (70,70), and each grid cell is 10 units in size in both x and y. Many applications use rectangular grids, and sometimes the data values are assumed to lie at the intersection points of the grid rather than at the grid squares. In this case, the grid squares are assumed to contain the data value.

To build a grid based on data at these points, we use the neighborhood property of the surface, that is, we assume that the elevations are continuously distributed, i.e., there

---

Function 11.01

```
File: data.xyz
Format: ASCII
        7          39          117
       27          17          122
       27          61          123
       46          36          136
       52          14          138
       55          46          141
       56          56          137
       65          24          144
       65          36          145
```

---

are no sudden cliffs, caves, overhangs, etc., and that values at any point are most closely related to elevations which are in the immediate proximity. Furthermore, we can assume that the influence of any point increases as distance to the point decreases, the so-called inverse-distance method. We can choose to make any interpolated surface fit exactly through the data points given, and we can choose whether to allow the highest and lowest points on the interpolated surface to fall beyond the ranges of the data at the points.

Many methods are available for interpolation. Some of them model the entire distribution of the surface and are considered in the *surface models* section which follows. The remainder are local operations, that is they work on small areas one at a time to achieve the interpolation. The first set of methods are the simplest, and use inverse distance as their theory. The methods work by moving from grid cell to grid cell, each time computing an interpolated value for that grid cell. In the following discussion, this grid cell will be termed the kernel.

11.2.1 Weighting Methods

Weighting methods work by assigning weights to the elevation values found within a given neighborhood of a kernel. The neighborhood is determined in one of several ways discussed in the following section on search methods. In a computer program, the search method computes the distance of each kernel to each point for every kernel. For a 200 by 200 grid, with 1,000 data points, this means 40 million distance calculations, each of which involves taking a square root and squaring two values, and then sorting each of the 1,000 distances to determine which is the closest. Hodgson (1989) called this the "brute force" method, and suggested an alternative "learned search" approach, in which the points are divided into coarser cells called sorted cells. The points are sorted within these areas and this allows the nearest point information to be retained as the algorithm moves from kernel to a neighboring kernel, saving considerably on the amount of time required.

Once the points within a neighborhood of a kernel are found, the distances from the kernel to each point are computed and used to inversely weight the elevations at the surrounding points. Mathematically, the formula used is

$$Z_{i,j} = \frac{\sum_{p=1}^{p=R} Z_p \, d^n_p}{\sum_{p=1}^{p=R} d^n_p} \qquad\qquad 11.01$$

where Z_p is the elevation at point p in the point neighborhood R, d is a distance from the kernel to point p, and Z[i,j] is the elevation at the kernel. The value n is the "friction of distance", which allows very distant points to be penalized with respect to closer points. As n increases, so also does the retention of breaks and extremes in the surface. When n is 1.0 and R is 3, the method is equivalent to linear interpolation over the TIN. Values of n for terrain have varied from 1.0 to 6.0, though many cartographers use a value of 2.0, in which case the technique is called inverse-squared distance weighting.

Refinements to the technique involve the insertion of barriers which the interpolator cannot cross, such as fault lines or coastlines, and using the cosine of the vertical angle between the point and the kernel in the weighting to eliminate the shadowing effect of closer points on the same bearing (Shepard, 1968).

Function 11.02 is a C language function which reads a file containing data points and creates a grid. The number of nearest neighbors is passed as an argument from the calling program. The neighborhood search, however, is performed unlike those described above. As points are read in, they are assigned their place in the as-yet-empty grid. If two or more points fall into one cell, they are averaged in. Each empty cell is then subjected to a search. Square neighborhoods around the cell are examined until the required number of points are found. It is possible that more than the required number can be found, since one extra row and column are searched on each side of the kernel in each scan. This algorithm has been found to be highly effective and is far faster than the brute force method.

Function 11.02

```
/*
 *
 * ========================================================
 * Inverse distance * squared gridding of x,y,z data
 * July 89      k.c.c.
 * ========================================================
 */
#include "header.h"
float               zsum, dsum;      /* Values computed externally */
#define NOTYET -9999    /* Value lower than lowest expected
                          elevation */
```

```
void inverse_d
(search_points)
    int             search_points;
{
    FILE            *pointsfile;
    int             k = 0, npts = 0, inc = 0, i, j, l, n, found, dd;
    char            filename[20];
    float           xmin = HUGE, ymin = HUGE, zmin = HUGE, x, y, z;
    float           xmax = -HUGE, ymax = -HUGE, zmax = -HUGE;
    unsigned short  empty[MAXROW][MAXCOL];
    /* Prompt for data file information */
    printf("\n Enter name of file containing data :");
    scanf("%s%*1c", filename);
    /* Open the data input file */
    if ((pointsfile = fopen(filename, "r")) == NULL) {
        printf("\n Error: Cannot open your input file \n"); exit();
    }
    printf("\n Data input in progress\n");
    /* Read the point data */
    while (fscanf(pointsfile, "%f%f%f%*1c", &x, &y, &z) != EOF) {
        npts++;
        if (x > xmax) xmax = x; if (y > ymax) ymax = y;
            if (z > zmax) zmax = z;
        if (x < xmin) xmin = x; if (y < ymin) ymin = y;
            if (z < zmin) zmin = z;
    }
    rewind(pointsfile);
    printf("  Minimum x value :%9f Maximum :%9f\n", xmin, xmax);
    printf("  Minimum y value :%9f Maximum :%9f\n", ymin, ymax);
    printf("  Minimum z value :%9f Maximum :%9f\n", zmin, zmax);
    printf("  Number of points processed : %d\n", npts);
    printf(" Boundaries of area to be gridded\n");
    printf("\n Enter the (x,y) for the lower left  :");
    scanf("%f%f%*1c", &grid.corners[0].x, &grid.corners[0].y);
    printf("\n Enter the (x,y) for the upper right :");
    scanf("%f%f%*1c", &grid.corners[2].x, &grid.corners[2].y);
    printf("\n Enter the desired number of columns (x)
        then rows (y)   :");
    scanf("%d%d%*1c", &grid.ncols, &grid.nrows);
    /* Echo the grid spacing */
    printf("One pixel is %f x units and %f y units\n",
        (grid.corners[2].x - grid.corners[0].x) /
        (float) grid.ncols, (grid.corners[2].y -
            grid.corners[0].y) / (float) grid.nrows);
    /* Initialize array */
    for (i = 0; i < grid.nrows; i++) { for (j = 0; j < grid.ncols;
        j++) {
            grid.z[i][j] = NOTYET; empty[i][j] = 0; }
```

```
    }
    /* Reopen and process the points file */
    printf(" Assigning points to grid\n");
    printf(" Grid is about %f percent full\n", 100.0 * npts /
        (grid.nrows * grid.ncols));
    for (n = 0; n < npts; n++) {
        fscanf(pointsfile, "%f%f%f%*1c", &x, &y, &z);
        i = (int) (grid.nrows * (float) (y - grid.corners[0].y) /
            (float) (grid.corners[2].y - grid.corners[0].y));
        j = (int) (grid.ncols * (float) (x - grid.corners[0].x) /
            (float) (grid.corners[2].x - grid.corners[0].x));
        if ((i >= 0) && (j >= 0) && (i < grid.nrows)
            && (j < grid.ncols)) {
            grid.z[i][j] += (int) z; empty[i][j]++;
        }
    }
    fclose(pointsfile);
    /* Adjust cells with more than one point */
    for (i = 0; i < grid.nrows; i++) { for (j = 0; j < grid.ncols;
        j++) {
            if (empty[i][j] > 0) { grid.z[i][j] -= NOTYET;
                grid.z[i][j] /= (float) empty[i][j];
                empty[i][j] = 0; } else empty[i][j] = 1; }
    }
    printf("Gridding begins..\n");
    for (i = 0; i < grid.nrows; i++) { if (!(i % 10))
        printf("Row %d0, i + 1);
        for (j = 0; j < grid.ncols; j++)
            if (empty[i][j]) { found = 0; inc = 0; dsum = 0.0;
                zsum = 0.0;
                while (found <= search_points) { inc++;
                    for (k = (i - inc); k <= (i + inc); k++)
                        { l = j - inc;
                        if (edges(k, l, grid.nrows,
                            grid.ncols) && !empty[k][l]) {
                            dd = (i - k) * (i - k) + (j - l)
                                * (j - l);
                            dsum += 1.0 / (float) dd;
                            zsum += (float) grid.z[k][l] /
                                (float) dd;
                            found++; } l = j + inc;
                        if (edges(k, l, grid.nrows, grid.ncols)
                            && !empty[k][l]) {
                            dd = (i - k) * (i - k) + (j - l)
                                * (j - l);
                            dsum += 1.0 / (float) dd;
                            zsum += (float) grid.z[k][l] /
                                (float) dd;
```

```
                                            found++; }
                                  }
                          for (l = (j - inc + 1); l <= (j +
                                inc - 1); l++) { k = i - inc;
                              if (edges(k, l, grid.nrows,
                                    grid.ncols) && !empty[k][l]) {
                                  dd = (i - k) * (i - k) + (j - l)
                                        * (j - l);
                                  dsum += 1.0 / (float) dd;
                                  zsum += (float) grid.z[k][l] /
                                        (float) dd;
                                  found++; } k = i + inc;
                              if (edges(k, l, grid.nrows,
                                    grid.ncols) && !empty[k][l]) {
                                  dd = (i - k) * (i - k) + (j - l)
                                        * (j - l);
                                  dsum += 1.0 / (float) dd;
                                  zsum += (float) grid.z[k][l] /
                                        (float) dd;
                                  found++; }
                          }
                  }
                      grid.z[i][j] = (int) (zsum / dsum);
                  }
          }
      return;
}
/* Detect edges of an array */
int edges(i, j)
      int               i, j;
{
      return ((i >= 0) && (j >= 0) && (j < grid.ncols) &&
            (i < grid.nrows));
}
```

11.2.2 Trend Projection Methods

Trend projection methods are designed to overcome the limitation that grid maxima and minima can lie only at data points. For a regularly sampled set of data points, it is rare that the grid will exactly coincide with the extreme highs and lows of the terrain. In trend projection, sets of points within a region are used. The region is usually determined using one of the search methods discussed in the next section. Fits to the surface are made locally, using a mathematical projection method such as a bicubic spline or a polynomial trend surface. This "best-fit" surface is then used to estimate the elevation at the kernel. If simple linear trend estimates are taken, often for triplets of points in the neighborhood, the actual value given to the kernel can be the mean of the various estimates, as in Figure 11.02, or the value can be weighted in the general direction of the surface trend.

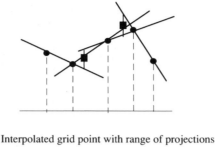

■ Interpolated grid point with range of projections

● Original data points (cross section)

Figure 11.02   Trend Projection Interpolation

In rough terrain, the trend projection methods may generate an interpolated surface which is far more textured than the true surface, but when the data are sparse and the need for texture information is high, this method may be useful. Sampson (1978) described in detail the trend projection method integrated into the SURFACE II package, a software system used extensively in the geosciences.

11.2.3 Search Patterns

Two major variables have a great influence on the two approaches to interpolation discussed above. The first is how the neighborhood is chosen. A neighborhood can consist of either a given number of points according to some criterion, or of all the points which satisfy certain conditions. The second influencing factor is the relationship between the spacing of the points and the spacing of the grid, and the related problem of the choice of the interpolation area and the orientation of the grid.

The simplest way of determining which points to use is to include all points in the interpolation. This method, while avoiding the sorting by distance described above, violates the neighborhood property nature of terrain data. Given that we must select points, one approach is to limit the search to points within a certain coarser resolution cell, determined by partitioning up the whole map. This leaves the problem of dealing with the discontinuities which result at the boundaries of cells. The brute-force method sorts all the distances, and takes for the neighborhood all points which fall less than a given distance or search radius away (Figure 11.03). This method can result in finding no or few points, especially at the map edges and corners. An alternative, which avoids this problem, is to take the nearest R points, regardless of distance or direction (Figure 11.04). Often, the nearest three, four, or eight points are selected. This method suffers when clusters of points exist, since any interpolation of kernels within the vicinity of the cluster will be overly influenced by the cluster. The problem of directional clustering can be overcome by choosing the nearest points within each of the four quadrants determined by the grid (Figure 11.05). Finally, each quadrant can be divided to assure equal representation of each octant (Figure 11.06). Both of these variations suffer from boundary effects; that is, at the edges and corners of the map grid, no points may be available within a quadrant of octant. This problem could be solved by moving to selecting the points for the neighborhood with another search method when no points are found in the search area. Clearly, some very complex search strategies can be constructed using combinations of these approaches.

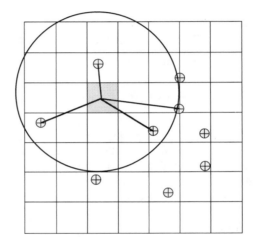

Figure 11.03   Interpolation Search by Radius

The second of the problems facing the interpolation search strategy is: How should the grid relate to the points? The orientation of the grid is usually the same as the coordinate system in use, although this is not a requirement, and for spherical data, where data should be interpolated across the poles, for example, some subtle refinements are necessary. The size of the grid is determined first by the spacing of the grid cells, that is, the grid resolution, and also by the area mapped. The total map area for interpolation is usually rectangular, and the map either is buffered by a data-point void strip around the edges, or extends over an area for which points lie beyond the map. Clearly, the latter is preferable, since values at the edges of the grid can be interpolated using real data outside the map area, giving the map reliability across the edges. When a void exists at the edge, some interpolator, especially the trend projection methods, can introduce artificial features into the terrain.

The grid spacing is also essential. At one extreme, a very large number of evenly distributed points and a coarse grid could allow the grid to be superimposed on the data and all cells to be assigned simply by averaging elevations within cells. At the other

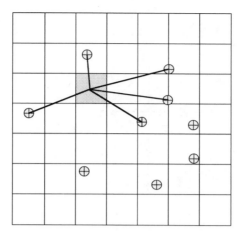

Figure 11.04   Interpolation Search by Number of Points

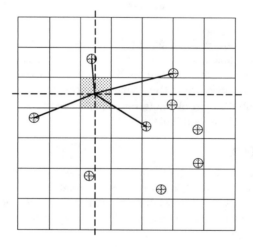

Figure 11.05   Interpolation Search by Quadrant

extreme, an extremely fine grid could ensure that every data point falls at a grid cell center exactly, leaving the problem of filling in the blanks. In fact, most irregular point distributions contain data-rich and data-poor areas. Statistics like the nearest-neighbor value may be helpful in determining grid spacing, as also is the mean, minimum, and maximum point separation. If the source of point data is field measurement, then an effort to get both a uniform spread of points and to collect the data extremes is critical. A map of the distance from each kernel to its nearest data point is a useful aid in determining where to draw the map boundaries and what grid spacing to use. No "best"

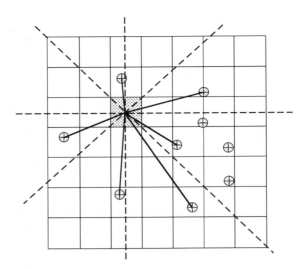

Figure 11.06   Interpolation Search by Octant

spacing exists, since the usual limitation is the collection of the data. The sampling theorem states that once a grid spacing is chosen, all features with a spatial size less than twice the spacing are essentially eliminated from the grid, at best becoming random variation or noise. Thus when a grid spacing is chosen, the purpose of the intended symbolization of the terrain should be carefully considered, so that features are not eliminated by smoothing.

11.2.4 Kriging

A very large number of methods have been used in the interpolation of point data to a grid. Only one method, however, uses statistical theory to optimize the interpolation. This method, advanced by Matheron and named for the originator, D. G. Krige, was originally applied to ore bodies in gold mining. Kriging is based on the mathematical theory of the *regionalized variable*. Regionalized variable theory breaks spatial variation down into a drift or structure, a random, but spatially correlated part, and random noise. Thus while hiking up a mountain, the elevation drift is up along a line between the trail head and the summit, even though we may find local drops to traverse along the trail (random correlated elevations), and boulders to step over while doing so (elevation noise).

Kriging involves a multi-step approach to interpolation. First, the drift is estimated using a mathematical function. If none exists, then a good estimate of any map elevation is the mean elevation of the data points. With drift, however, the expected elevation difference with a given separation between the kernel and a point is given by the semi-variogram. The semi-variogram is computed within regions determined using one of the search strategies discussed above. The semi-variogram is then used to statistically fit a model, usually an exponential but a linear model when the semi-variogram has no obvious sill, to the distribution. This allows the estimation of semi-variance, and also the estimation of the weights to be used in computing the local moving average. The weights are chosen so that the best unbiased values are used, and the estimation variance is minimized. A more detailed discussion, with numerical examples, is contained in Burrough (1986).

Since kriging yields a surface which passes directly though the data points, and since the technique also yields the estimation variance at each interpolated point, the technique is statistically superior to the interpolation methods discussed above. The method is available in several computer packages, even on microcomputers, for example the SURFER package. The technique of universal kriging works best for data with well-defined local trends. When it is difficult to use the points in a neighborhood to estimate the form of the semi-variogram, the model used is not entirely appropriate, and the interpolation may be no better than another method. While kriging is optimal in statistical terms, it is very computationally intensive, and can take up a great deal of computer time, especially for large numbers of data points and large grids. Burrough (1986) has noted few comparative studies of the results of kriging compared to other methods. Figure 11.07 shows the results of gridding the test point data set presented above, first using inverse distance squared weighting with a four-point nearest-neighbor search, and second using universal kriging. In the figure, areas of discrepancy between the two methods are shaded. In only one small area are the kriging estimates significantly higher than those of the distance weighting—in the data-poor zone of the map corner, beyond a point cluster. In other areas, the discrepancies in the distance weighting are overestimates, with the corners, edges and sparse data areas predominating. Clearly, in such a small grid, boundary effects are major, and seem to hold the balance between differences in the two interpolations.

Research into interpolation continues, and algorithms for gridding are becoming increasingly sophisticated. The major source of data to be interpolated, in addition to field or survey data, is data digitized from contours. Contours are a special case, since they are already graphic interpolations of the point data. In many cases, contour lines contain more information about the terrain than simply the elevations, since when drafting the map, the cartographer often traced features detected on the ground or drew the contours to show streams. Digitizing points from contour maps, therefore, is complex. Simply scanning the contour separation of a map is usually a poor way of generating digital terrain. A more effective approach is to use surface-specific point sampling and to then use either a TIN or an interpolated grid with known interpolation properties to map the terrain.

11.3 SURFACE-SPECIFIC POINT SAMPLING

The logic behind surface-specific point sampling is that contour maps, and even field elevation data collected by topographic surveys, contain information about the terrain surface which would be lost by placing a grid over the map and digitizing elevations at grid cells. Significant features which dominate the form of the terrain are streams, ridges, summits, saddle points, and the bottoms of depressions (Douglas, 1986). In

125	124	123	127	135	137	138
121	125	125	131	136	137	139
118	120	128	135	138	141	141
117	119	127	135	136	141	145
118	123	125	131	137	140	144
123	123	122	128	136	138	141
125	124	123	129	135	138	140

119	121	124	128	133	138	141
118	120	122	128	135	140	143
117	119	122	128	136	142	145
116	119	123	129	136	143	145
117	119	123	129	136	142	144
118	119	122	127	133	137	141
119	120	123	127	131	135	138

Inverse Squared Distance Weighting
with 4 point nearest neighbor search

Universal Kriging

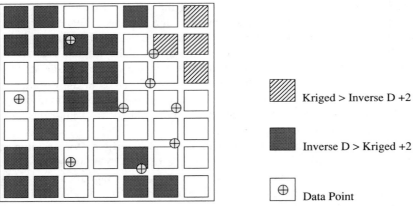

Kriged > Inverse D +2

Inverse D > Kriged +2

⊕ Data Point

Figure 11.07   Comparison of Two Interpolations for Test Data

addition, actual point elevations are often available on the map as benchmarks, data collection points, or spot heights.

Ignoring the structure or "skeleton" of the terrain, one approach to converting a contour map into either a grid or a TIN is to digitize points along the contours with the attribute of the elevation of the contour. This may work well in only two circumstances: when the point density along the lines is about the same as the map spacing between the contours, and in terrain which is very rough. In most cases, more points are digitized along the contours than between them, as shown in Figure 11.08. When this is the case, interpolation, especially when only a few points are used in the point search, will result in artificial plateaus within the loops of the contours. Since contours usually move back and forth much like a stream, the result is a series of flat terrain steps along the slope line. These steps are artificial and are the result of poor digitizing technique associated with the weaknesses of certain interpolation methods.

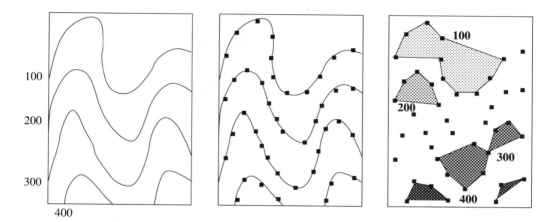

Figure 11.08   Errors in Gridding from Digitized Contours

Avoidance of this problem means taking into account the skeleton of the terrain. Digitizing software which permits an automatic increment by the contour interval with each point digitized is particularly valuable when digitizing streams and ridges. An effective strategy is to start at the low point on a stream or ridge, and to continue up until the high point is reached. Saddle points, summits, and pits (depressions) are then places where the stream and ridge lines converge (Figure 11.09). Additional ridges can follow significant features up slopes, such as the ridge formed by a stream incision. The significant points, such as summits themselves, and other benchmarks, should also be entered. Finally, the intermediate empty spaces can be filled with the occasional point digitized along contours, although following a single contour should be avoided. The result is a set of points which best represent the surface of the terrain. If the TIN were generated from these points, the TIN would be a best-fit model of the surface, and the triangles generated would be good representations of the actual slope facets on the ground. This set of points is also optimal for interpolation, since the interpolator is often better at assigning intermediate elevations from the skeleton than the original manual cartographer was at drafting contours.

11.4 SURFACE MODELS

The interpolation methods discussed so far share the fact that they fit the interpolated surface by local adjustments. Other methods, especially those which for analytical or modeling purposes seek to generalize the surface, use a global or entire surface approach. The *surface models* fall into categories based on the mathematics which

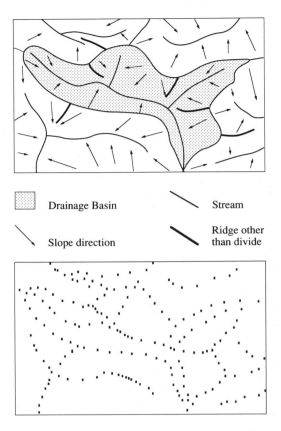

Figure 11.09   Surface-Specific Point Selection

describe the surface. The two major surface models, which are global in scope, are polynomial series and Fourier series. Each has analytical powers beyond generalization, and each has been used extensively in analytical cartography.

11.4.1 Polynomial Series

Polynomial series are power series, in that they are series summations of increasing powers of eastings and northings and their cross products. In the simplest form, a linear surface can be represented mathematically as

$$Z_{x,y} = \beta_0 + \beta_1 X + \beta_2 Y \qquad\qquad 11.02$$

As the first constant increases, so the overall average height of the surface increases. As each of the others increases, so the linear plane surface represented by the equation dips or tilts down more and more up to the west or south, respectively. Since a flat tilting plane is rare as terrain, real data seldom falls on such a plane, but overshoots and undershoots it. Figure 11.10 shows a linear surface passing through the point data set used earlier in this chapter. Such a surface can be thought of as the overall trend in the data, the drift or trail head-to-summit path discussed above. On top of this trend are the local variations, the overshoots and undershoots. These are called residuals, and overshoots are positive residuals, while undershoots are negative. Statistically, the residuals can be used to fit equation 11.02 to a set of points or a surface using least squares. The technique computes the beta coefficients above which minimize the total of the residuals squared, since some are positive and some negative. The resultant trend therefore has an associated "goodness of fit" measure, which gives the percentage of the variance in the elevation values accounted for by the linear trend model. For the data in the test data set, a linear trend was fit to the inverse-distance-squared weighted interpolation with four points shown in Figure 11.11. The resultant trend has beta coefficients of 119.66, 3.73, and -0.0310 respectively, with a figure of 79.23 % of the surface variance explained. The negative coefficient for the northing implies that the slope dips very gently to the north. The slope to the east is steeper, and the first constant is the

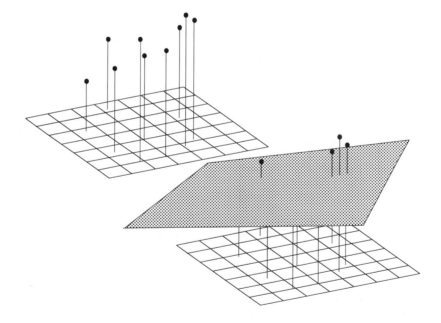

Figure 11.10   A Linear Trend Surface

value at location (0,0). Figure 11.12 shows a map of this trend surface. More valuable than the trend, however, is a map of the residuals from the trend (Figure 11.13). The mapping of residuals has found many applications in geology and geography (Davis, 1973). When residuals cluster, usually a larger scale spatial process is at work. Increasing the order of the polynomial is one way to capture this variation in the trend

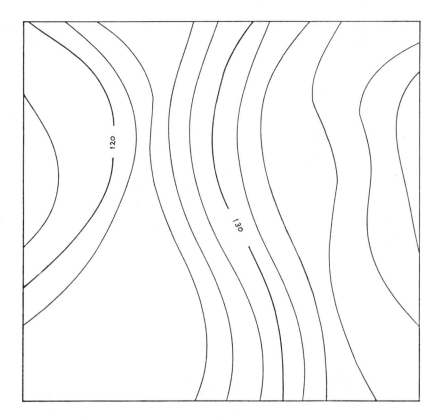

Figure 11.11   Test Data Set Gridded Using Inverse-Distance-Squared Weighting and
Four Nearest-Point Search

surface model. The least squares principle is applied to higher-order polynomials. For
example, a quadratic surface would be given by

$$Z_{x,y} = \beta_0 + \beta_1 X + \beta_2 Y + \beta_3 X^2 + \beta_4 Y^2 + \beta_5 XY \qquad 11.03$$

As the surface becomes more complex, so the number of beta terms increases. A cubic
surface has 10 terms, more than the number of points from which our test data set was
derived. At this point the value of the trend surface as a generalization of the original
data becomes questionable, since the model has more parameters than we have data.
For mapping purposes, such as generalizing terrain for mapping, or for simplifying
slope and aspect computations, a trend surface of appropriate order is adequate. As a
theoretical model of terrain, however, the trend surface may introduce unacceptable
error during generalization. Davis (1973) provided an excellent discussion of the use
and theory of trend surface analysis in geology, and published a FORTRAN computer
program for its implementation.

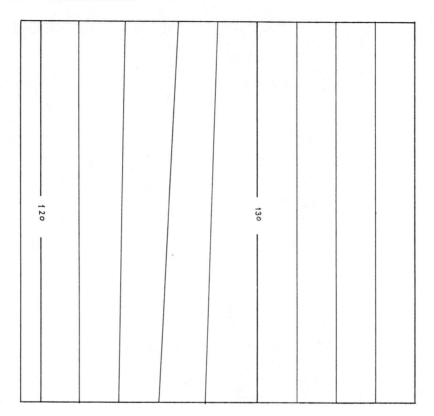

Figure 11.12   Linear Trend Surface for the Test Data

11.4.2 Fourier Series

An alternative method for the generalization of terrain is to use a theory developed by the French mathematician Fourier. Fourier showed that any sequence of data could be represented by the sum of a set of trigonometric functions, sines and cosines, over a long enough range of wavelengths. Using space rather than time, since Fourier methods are most commonly used for time series analysis, the wavelength is the distance at which the wave repeats itself. The amplitude of the wave for terrain is the elevation, and the phase angle, the point at which the wave starts, is the elevation at distance zero.

A Fourier generalization of terrain works in two dimensions, x and y, and gives values for elevation. To compute the parameters of the trigonometric functions, Fourier coefficients are computed for all ranges possible within a data set. Since Fourier analysis of irregular data is complex, the usual data used for Fourier analysis is the grid. Analysis proceeds starting at the longest-wavelength wave that will fit the data, that is, with a wavelength of half the map length in x and half the map length in y. The

Figure 11.13   Residuals from the Linear Trend Surface

response or "power" of each pair of x and y wavelengths is computed, a value similar to the percentage of variance explained in least squares. The wavelengths used are the map length L over 2, 3, 4, 5, etc., until the distance associated with the wave approaches twice the grid spacing, a level at which the Gibbs phenomenon introduces seemingly random errors. This gives a new grid of Fourier coefficients, which are normalized to give the "power" of each pair of wavelengths or "harmonics".

The generalization involves selecting particular pairs of harmonics which are major contributors to the structure of the data. In most phenomena, including linear phenomena such as voice and radio transmission, just a few of the wavelengths carry almost all of the structure of the data. These harmonic pairs can be extracted and their Fourier coefficients used to reconstruct a data grid based on their values. This is another example of an inverse transformation. The conversion from the spatial to the frequency or wavelength domain is known as the Fourier transform. The inverse Fourier transform can be used to reconstruct either the entire original data grid (less some errors due to boundary problems), or part of it. A common use of the Fourier transform and its inverse is to eliminate the higher frequencies, that is, the variations at small distances. The forward and inverse Fourier transforms give an objective method for this form of generalization.

Davis (1973) gave a FORTRAN program for performing two-dimensional Fourier analysis. Davis's method used the discrete Fourier transform, a grid cell-by-grid cell method. Far faster is the fast Fourier transform. However, this technique requires data sets to be of specific sizes. Algorithms, equations, and C language computer programs for both forms of the Fourier transform are contained in Press et al. (1988). Application of the discrete Fourier technique to terrain is discussed in Clarke (1988). The by-product of Fourier analysis of terrain is a precise account of which spatial scales are "active" in a particular piece of terrain. This assists in the choice of a sampling grid, as well as in analyzing the processes which have formed the terrain surface.

11.4.3 Surface Filtering

An effect virtually identical to Fourier generalization is possible using local operators. This method is called spatial filtering. Filtering is usually performed exclusively on gridded data, and the TIN will therefore be left out of the following discussion. In spatial filtering, the entire grid is processed cell by cell to generate a new or filtered grid. This is done by using a smaller grid, and moving the filter grid step by step, kernel by kernel, over the original data grid. In each case, the filter grid is used to compute a moving average. Filter grids are centered on a kernel, and must therefore have an odd number of rows and columns. Filters of 3-by-3, 5-by-5, and 7-by-7 are common. Larger filters lead to problems, since in each application of the filter the original grid gets smaller. Three-by-three filters lose one row from each edge of the grid with each application, five-by-five lose two, etc. The filter is placed over the original data grid, and its values are multiplied by the originals to give new values, which are then summed. The new filtered value for the grid is then the sum of the values over the filter (Figure 11.14). A requirement for the filter is that the weights or values within the filter grid sum to 1. Otherwise, the filter has the effect of damping (< 1.0) or amplifying (> 1.0) the terrain. Filter weights can be tailormade to yield different effects. Equal weights in each of the cells is simply a moving average or smoothing filter. The weights can be distance weighted by weighting the cells in proportion to their distance from the kernel. A common filter is the Hanning filter, a two-dimensional version of the binomial distribution.

Special purpose filters can be designed to enhance features with a specific size, orientation, or characteristic. A horizontal linear filter, for example, can be used to amplify erroneous scan lines on satellite images. A filter with the shape of a ship, for example, can be used to scan ocean images for ships. Another commonly used filter is the "lint-picker", a 3 by 3 filter with a weight of -1 in each cell except the center, which has a weight of 9. Note that the weights in this filter still sum to 1. The effect of this filter is to enhance features which are one grid cell in size and are very different from

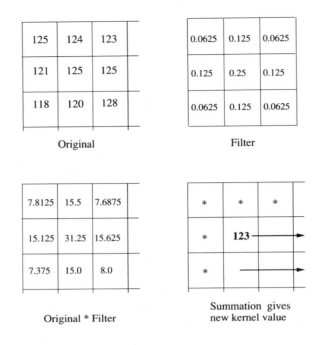

125	124	123
121	125	125
118	120	128

Original

0.0625	0.125	0.0625
0.125	0.25	0.125
0.0625	0.125	0.0625

Filter

7.8125	15.5	7.6875
15.125	31.25	15.625
7.375	15.0	8.0

Original * Filter

*	*	*
*	**123**	
*		

Summation gives
new kernel value

Figure 11.14   Application of a Spatial Filter

their neighbors. This is a good definition of random noise, so a lint-picker is often used to generate an image or map which is subtracted from the original to remove random noise, the so-called "computer-enhanced image". Filtering an image is fairly easy to program, but can be computationally intensive, since the whole grid must be used and a second version of the grid must be saved in RAM during processing.

Function 11.03 is a C function which filters a grid in the GRID structure. The weights are stored in a vector, and can be adjusted for any size or type of filter with a little modification. The function shown assumes a 3 by 3 window and a Hanning filter.

11.5 VOLUMETRIC CARTOGRAPHIC TRANSFORMATIONS

Surface models and filtering are transformations of the three-dimensional cartographic data. So far, the motive for transformations discussed have been to transform between data structures (interpolation) and to transform between scales (generalization). Many other transformations are commonly applied to digital elevation data. The purpose of these transformations is purely analytical; that is, the result is a map with enhanced meaning for a specific cartographic problem. Four aspects of analytical transformations will be discussed in this section: the transformation of elevations to slope and aspect, the automatic delineation of terrain-significant points from gridded data, the simulation of terrain data, and the transformations necessary to provide visibility maps.

Function 11.03

```
/* ======================
/* Filter a grid
/* July 1989 kcc
/* ====================== */
#include "header.h"
filter() {
    static float    filter_weights[9] = {
    0.0625, 0.125, 0.0625, 0.125, 0.25, 0.125, 0.0625, 0.125,
         0.0625 };
    float           filtered_array[MAXROW - 2][MAXCOL - 2];
    int i, j, l, m, index;
    for (i = 1; i < (grid.nrows - 1); i++) {
        for (j = 1; j < (grid.ncols - 1); j++) {
            index = 0;
            for (l = i - 1; l <= i + 1; l++) {
                for (m = j - 1; m <= j + 1; m++)
                    filtered_array[i - 1][j - 1] +=
                        grid.z[l][m] * filter_weights[index++];
            }
        }
    }
    grid.nrows -= 2; grid.ncols -= 2;
    for (i = 0; i < grid.nrows; i++) for (j = 0; j < grid.ncols;
         j++)
        grid.z[i][j] = filtered_array[i][j];
    return;
}
```

11.5.1 Slope and Aspect

Evans (1980) noted that slope is defined by a plane tangent to a surface at a specific point, and is specified in terms of the maximum rate of change of altitude (slope) and the compass direction associated with the maximum (aspect). Slope, therefore, also has a local neighborhood over which it is computed, usually a 3 by 3 region. The maximum slope is familiar to skiers as the fall line, and the aspect is the direction in which the fall line trends. When slope is zero, the terrain is flat, and the aspect is undefined. Slope is computed by solving a best-fit surface through the points in the neighborhood, and by measuring the change in elevation per unit distance in this neighborhood, and the direction. These values can be assigned as data to a new grid. Using the TIN, each triangle has a uniform slope within the triangle, and also a single aspect. The slope and aspect values in a TIN are discontinuous at the triangle boundaries. Figure 11.15 shows values of slope and aspect computed for an elevation grid covering the summit of Mount Everest. The horizontal resolution is 18 meters per grid cell. The slope map shows a limited set of slopes with extremely steep slopes in white. The aspect image shows with gray tones the eight major compass directions, starting with black for north and ending with white for north west.

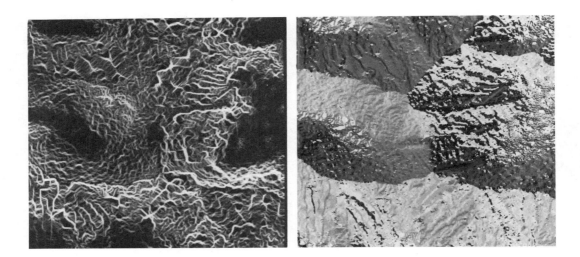

Figure 11.15   Slope and Aspect Maps for the Everest Summit Area

11.5.2 Terrain Partitioning

The primary data structure conversion for three-dimensional data, with the exception of the point to grid or point-to-TIN conversion, is the TIN-to-grid, and vice versa. A TIN-to-grid conversion is comparatively simple. Either the points in the TIN can be used in a grid interpolation using one of the methods discussed in this chapter, or the elevation values at every point in the grid can be computed using a linear, polynomial, or spline fit to the data in the triangle which includes the point. The grid-to-TIN conversion, however, is more difficult. There are distinct advantages to performing this conversion. TINs are very compact, are suited to symbolization using polygon rendering hardware and software, and are computationally less demanding than grids. Essential to the advantages of the conversion, however, is the ability of an algorithm to detect critical points in the landscape. These are the same points discussed as surface-specific in the preceding section. A factor of reduction of 18 times was achieved for conversion of triangles formed by grid points to a TIN based on critical points (Scarlatos, 1989).

The detection of stream and ridge lines in gridded terrain data was discussed by Douglas (1986) Peaks, saddles, and pits are detectable by filtering. The advantages of detecting the significant points is not only for creating the TIN, but also for automatically detecting river channels, selecting ridges for hill shading and intervisibility, and for dividing the surface up into sloping facets for the modeling of hydrology. As yet, however, no simple algorithm to convert a grid to a TIN exists.

11.5.3 Terrain Simulation

Interest has recently been focused on the simulation of artificial terrain, with realistic characteristics. Such terrain is used in flight simulators, movies, games, and in models of natural processes. The most common means by which such terrain is produced is using the mathematical methods of fractal geometry. Fractal geometry defines a property called self-similarity, in which as the scale gets larger and larger, geometric form is simply repeated. Two methods for the generation of fractal terrain are commonly used. The first is the midpoint displacement method, in which an original square is drawn, its four corners displaced either up or down at random, and then the square is divided into four quarters, whose corners are displaced by the same amounts, and so on. In each case, the center cell is the average of the four corner cells. This method was first used by Fournier, Fussell, and Carpenter (1982) for animations, and is fast, simple to program, and gives satisfactory results. For many divisions, however, the technique leaves creases at the edges of the larger cells, for which Jeffery (1987) provided a solution.

A second algorithm, given in Jeffery and attributed to Voss, adds the displacements to all points at each scale instead of just the mid-points. Since this allows steps down of more than a half at a time, the creasing problem disappears. Jeffery (1987) published pseudo-code for his algorithms, and Pascal versions of the programs are available. Information on accessing the programs is contained in an editor's note in the article. While many fractal surfaces are perfectly adequate for simulations, cartographically incorrect characteristics arise. For example, as many pits as peaks are generated. Often, additional texture algorithms or filtering are used to produce the final "fractal forgery".

11.5.4 Intervisibility

The intervisibility problem can be stated as follows. Given a digital cartographic representation of a terrain surface and a single point on or above the surface, determine the set of regions on the surface which are visible from that point. This set of regions is the opposite of the set of invisible regions. A map of either visible or invisible regions has immense value. Intervisibility maps are used in siting radar and television transmitters, in locating fire towers, in planning ski resorts and housing developments, in highway planning, and by the military. Several related problems, such as planning sets of points within view of each other, hiding environmentally obtrusive buildings and land uses in the terrain, and computing the regions where an aircraft is visible to radar, are of direct use in many areas.

The simplest approach to determining intervisibility is to connect a viewing location to each possible target and to follow the line back looking for points which are higher, a method known as ray tracing. A higher point would screen the target from the viewer (Figure 11.16). Since this involves a very large number of computations, the screening out of invisible areas along rays is advantageous. Intervisibility is easier using the TIN than on a grid. Sutherland et al. (1974) reviewed several algorithms, while DeFloriani et al. (1986) provided a TIN algorithm and Anderson (1982) a grid algorithm. An improved method for the grid was published by Dozier et al. (1981).

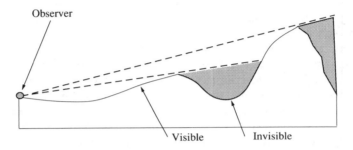

Figure 11.16   Intervisibility Using Ray Tracing

A primary use of intervisibility algorithms is in selecting viewing locations for gridded and realistic perspectives. The algorithms are also used in calculating the hidden sections of perspective views, and in eliminating hidden lines from gridded perspective views. Figure 11.17 shows an application where rays are traced from a viewpoint and are used to assist the cartographer in the selection of a viewing position for generating a realistic perspective view. This step can act as a preprocessor for this computationally intensive task.

11.6 TERRAIN SYMBOLIZATION

The final cartographic transformation for cartographic data is that which produces a map. Maps of three-dimensional data have problems similar to those with map projections, that is, how to produce on a flat two-dimensional plane a map depicting a volume in three-dimensions. Several cartographic techniques have been devised in recent history for the cartographic symbolization of three-dimensional data. The definitive study of manual methods is Imhof (1982). These methods include contouring and hill shading, as well as block diagrams. With computer cartography, many additional methods have arisen, and many methods which required much labor by hand have become available to all.

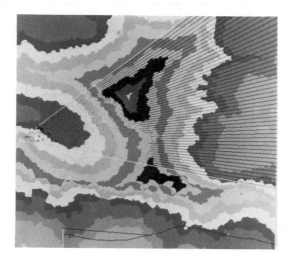

Figure 11.17   Ray Tracing for Viewpoint Selection

11.6.1 Automated Contouring

The oldest non-pictorial method of representing three-dimensional data is to use the isoline, first used to show magnetic variation by Sir Edmund Halley. When applied to terrain, the isoline is called a contour, and is a line joining points with equal elevation. Reference contours are drawn thicker and numbered at straighter segments, and closed depressions are annotated with hatch marks to distinguish them from hills. Few computer contouring programs automate these features, however.

Computer contouring can be performed from either a TIN or a grid. Starting either at one side of the map, the highest elevation, or the lowest elevation, a first step is to determine whether or not a contour line should appear in a given grid cell or triangle. If the answer is yes, the next step is to determine points at the edge of the square or triangle where the contour enters and leaves. The next stage uses an interpolator either a quadratic trend surface, an average slope, or Lagrangian interpolation (Crain, 1970)—to generate points along a curve within either the square or the triangle. One exception in the grid cell case complicates this process, the case of a saddle point (Figure 11.18). In this case, a center point is computed as the average of the four corners and used to move the contour to one side of the cell center. The final contour lines are then smoothed, using weighted averaging or spline functions. A final map consists of a set of points for each contour still structured by grid cell or TIN triangle. In an effort to assure continuity, the contour lines are often resorted so that each continuous loop is drawn without breaking the line, a step which improves plotter output considerably but is unnecessary for many output devices.

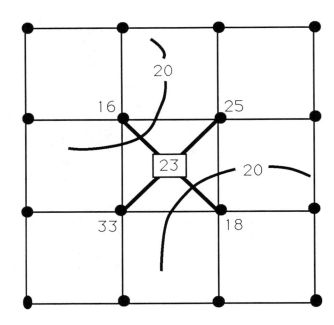

Figure 11.18   The Saddle Point Problem in Automated Contouring

As an example, Figure 11.19 is an automated contour map of the same data as Figure 11.11 with the exception that the generated contours were smoothed using a different tension factor for the spline smoothing. In some cases, using incorrect values for these coefficients can lead to erroneous contour lines. Some contours become dots at peaks, some cross, and some converge as lines. The latter feature is visible on figure 11.20, an automated contouring of the Everest summit data used above for slope and aspect computations. Several inexpensive packages now permit rapid contouring even on microcomputers. It is important to remember, however, just how many parameters go into the production of a computer-generated contour map. The choice of these parameters, data structures, and methods, not to mention the type of output device, are strong determinants of the look of the final contour map. In some cases, the only difference between a manual and a computer contour map is the computer's ability to reproduce the same map given the original data and parameters used.

11.6.2 Analytical Hill Shading

A terrain representation method which has gained considerably in use with computer cartography is hill shading. Automated hill shading works by computing the three-dimensional vector normal to the surface at each point in a grid, or for each triangle in a TIN. This normal vector is at right angles to the plane which defined the maximum slope, and faces in the direction of the aspect. This vector is projected in three-

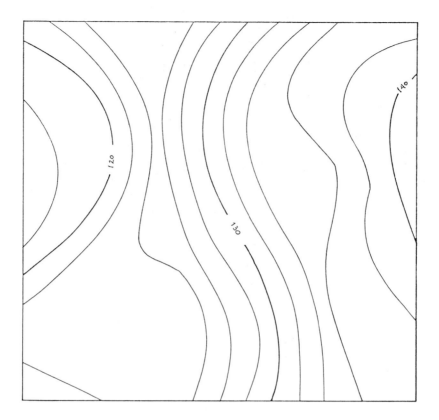

Figure 11.19   Figure 11.11 with Different Spline Coefficients

dimensions onto the vector given by a simulated sun angle. The sun, or light source, since multiple sources are possible, is located off the map, with a given zenith and azimuth. The zenith is the vertical angle of the light from the center of the map, while the azimuth is the compass bearing of the sun's direction. If the normal vector faces away from the sun, the surface at that point will be in shadow. If the normal vector points directly to the sun, the point will be fully illuminated. At other angles between the illumination and the normal vector, the point will be partially illuminated. Analytical hill shading, first derived by Pinhas Yoeli, records the illumination value across the map, usually for every grid cell in a grid, but also over a TIN. Brassel (1974) noted that two values can be used, a reflectance value as discussed above, or a density value, which is the logarithm of the inverse reflectance. Figure 11.21 shows the Everest summit elevation data with hill shading using four different solar illumination directions, NE, SE, NW, SW, and solar azimuths of 40 degrees. With control over these angles, it is possible to generate impossible illumination as far as nature is concerned. Hill shading is often used to modify color information on maps derived from remote sensing to

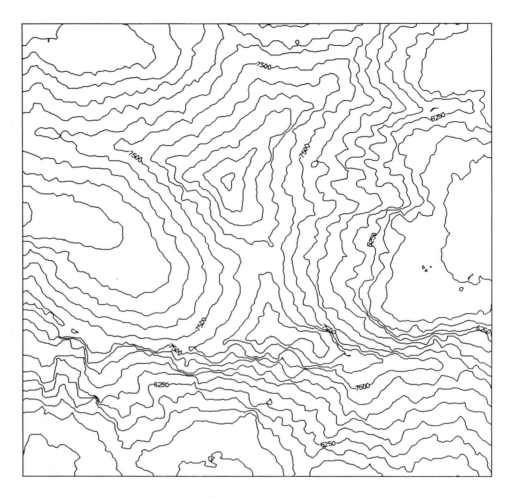

Figure 11.20   Automated Contour Map of the Everest Summit Region

present more striking images. Hill shading is also important for generating realistic perspective views.

11.6.3 Gridded Perspectives

The first computer cartographic equivalent of the block diagram used so commonly in geology was the gridded perspective map. These maps are produced exclusively from gridded data, since TINs viewed in perspective appear too complex. Tobler (1970) published a FORTRAN program to generate these views, without hidden-line processing. Many software vendors now sell computer programs to produce gridded perspectives, even on microcomputers with menu control. The critical values in the generation of

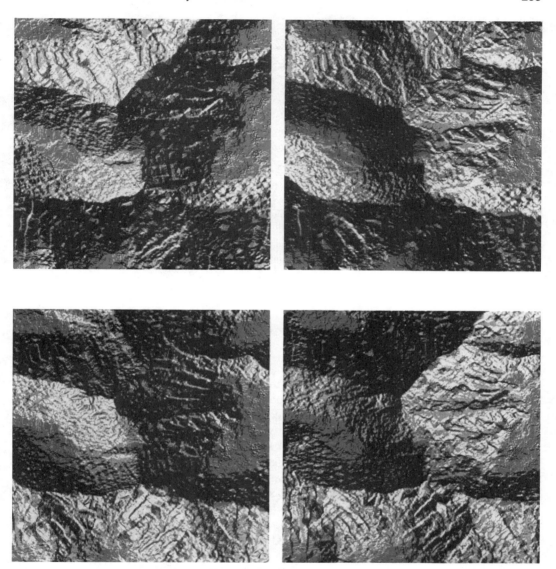

Figure 11.21   Reflectance Hill Shading with Different Illumination

these images are the viewpoint, the vertical exaggeration, the skirt, the alignment of the lines, and the addition of scales (Figure 11.22). The viewpoint establishes several things, including the perspective. Views from too close have a "wide-angle" perspective, while distant views produce the orthographic perspective favored in block diagrams. Viewpoint can be specified by azimuth, zenith, and distance, by locating a

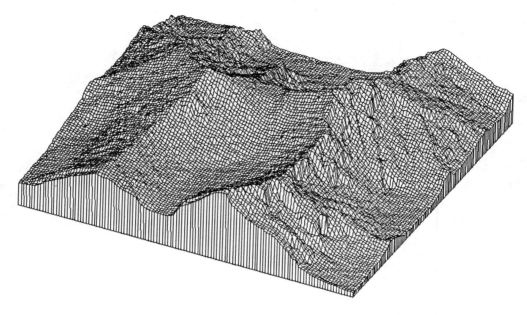

Figure 11.22   Gridded Perspective View of the Everest Summit Area

center of the field of view, or by giving a precise triplet of coordinates. The geometry is usually with respect to the center of the volume represented by the image.

The vertical exaggeration is important to the look of the perspective. Textured terrain, when exaggerated too much, looks chaotic, while too little terrain exaggeration cloaks any actual relief. The skirt is the plain base of the figure. Typically, control over the elevation of the base, as well as whether the surface lines will be drawn over the base, is available. Without a skirt, gridded perspectives seem to float, but if parts of the underside are visible, added information is gained, especially if the underside is colored differently. Lines on the perspective are usually square to the x and y axes, but variants are to make the lines parallel to the line of sight, aligned to the z axis (raised contours), or sometimes any combination, perhaps in different colors.

Two variants are stereo plots and anaglyphs, both means by which stereo views can be simulated. In a stereo plot, two images are generated, separated by a 2-degree viewing difference, and at a spacing suitable for use with a stereo viewer (Figure 11.23). The geometry of this view is also shown in Figure 11.23, which can be viewed in stereo using a pocket stereoscope. Finally, the anaglyph plot is identical, with the exception that the two different views, with a 2 degree viewpoint separation, are plotted on top of each other. One image is plotted in green or blue, and the other in red. Viewing the image through anaglyphic glasses, which have red and blue or red and green lenses, then, produces the stereo effect. The stereo image, combining two color opposites, should appear as black.

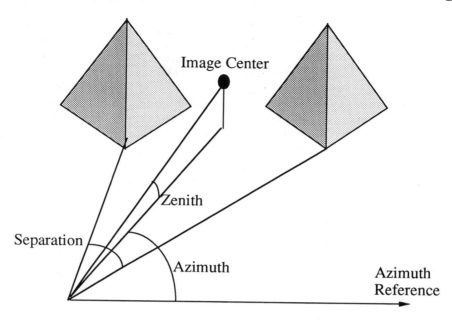

Figure 11.23   Stereo Gridded Perspective View of Test Data

11.6.4 Realistic Perspectives

A refinement of the gridded perspective is the realistic perspective. Dubayah and Dozier (1986) presented a summary of work on this method, and discussed algorithms. These images can be generated from both grids and TINs, but employing very different methods. TINs are usually rendered using special purpose hardware or display software, whereas grids are usually processed in batch mode, and the image displayed after completion. The more powerful workstations are capable of almost real-time generation of these images, but even powerful microcomputers may take hours of processing to produce a single image. Hours of supercomputer time have been used to produce sets of these images, which can then be played back in sequence to simulate flight and motion. Most of the same parameters as gridded perspectives apply to realistic perspectives. One major problem is the enlarged effect of grid cells very close to the observer, which can appear blocky. Color for these images is often natural color derived from satellite data. The cover of this book was generated by integrating hill shading with a color sequence based on elevation. The view is from the west, with a viewpoint at the edge of the Everest map used throughout this section, just slightly above the terrain surface.

11.7 TRANSFORMATIONS IN REVIEW

This section began with a discussion of the cartographic transformation. We have seen that a transformational view of cartography is broad enough to encompass all of both computer and analytical cartography. While analytical cartography represents the intellectual challenge to the cartographer, it is the act of producing the maps which often takes up much of the digital cartographer's time. In the following section of the book, an approach is presented which ensures that this expenditure of time is efficient. To incorporate cartographic transformations as algorithms into mapping systems demands that the cartographer understand the art and science of computer programming. Both analytical and computer cartography have much to gain from effective computer programs, just as cartography as a whole can benefit from better and more available maps.

11.8 REFERENCES

ANDERSON, D. P. (1982) "Hidden line elimination in projected grid surfaces", *ACM Transactions, Graphics*, vol. 1, no. 4, pp. 274-291.

BRASSEL, K. E. (1974) "A model for automatic hill shading", *The American Cartographer*, vol. 1, no. 1, pp. 15-27.

BURROUGH, P. A. (1986) *Principles of Geographical Informations Systems for Land Resources Assessment*, Clarendon Press, Oxford.

CLARKE, K. C. (1988) "Scale-based simulation of topographic relief", *The American Cartographer*, vol. 15, no. 2, pp. 173-181.

CRAIN, I. K. (1970) "Computer interpolation and contouring of two-dimensional data: a review", *Geoexploration*, vol. 8, pp. 71-86.

DAVIS, J. C. (1973) *Statistics and Data Analysis in Geology*, Wiley, New York.

DEFLORIANI, L., FALCIDIENO, B. AND PIENOVI, C. (1986) "A visibility-based model for terrain features", *Proceedings, Second International Symposium on Spatial Data Handling*, International Geographical Union Commission on Geographic Data Sensing and Processing and the International Cartographic Association, Seattle, WA, July 5-10, pp. 235-250.

DOUGLAS, D. H. (1986) "Experiments to locate ridges and channels to create a new type of digital elevation model", *Cartographica*, vol. 23, no. 4, pp. 29-61.

DOZIER, J., BRUNO, J. AND DOWNEY, P. (1981) "A faster solution to the horizon problem", *Computers and Geosciences*, vol. 7, no. 2, pp 145-151.

DUBAYAH, R. O. AND DOZIER, J. (1986) "Orthographic terrain views using data derived from digital elevation models", *Photogrammetric Engineering and Remote Sensing*, vol. 52, no. 4, pp. 509-518.

EVANS, I. S. (1980) "An integrated system of terrain analysis and slope mapping", *Zeitschrift fur Geomorphologie*, Supplement -B.d. 36, pp. 274-295.

FOURNIER, A., FUSSELL, D. AND CARPENTER, L. "Computer rendering of stochastic models", *Communications of the ACM*, vol. 25, no. 6, pp. 371-384.

HODGSON, M. E. (1989) "Searching methods for rapid grid interpolation", *The Professional Geographer*, vol. 41, no. 1, pp. 51-61.

IMHOF, E. (1982) *Cartographic Relief Presentation*, DeGruyter, New York.

JEFFERY, T. (1987) "Mimicking mountains", *Byte*, December, pp. 337-344.

PRESS, W. H., FLANNERY, B. P., TEUKOLSKY, S. A. AND VETTERLING, W. T. (1988) *Numerical Recipes in C : The Art of Scientific Computing*, Cambridge University Press, New York.

SAMPSON, D. (1978) *Surface II Graphics System*, Kansas Geological Survey, Lawrence, KS.

SCARLATOS, L. L. (1989) "A compact terrain model based on critical topographic features", *Proceedings, AUTOCARTO 9*, Ninth International Symposium on Computer-Assisted Cartography, Baltimore, MD, April 2-7, pp. 146-155.

SHEPARD, D. (1968) "A two-dimensional interpolation function for irregularly spaced data", *Proceedings*, Twenty-third National Conference, ACM, pp. 517-524.

SUTHERLAND, I. E., SPROULL, R. F. AND SCUMACKER, R. A. (1974) "A characterization of ten hidden-surface algorithms", *Computing Surveys*, vol. 6, no. 1, pp. 1-55.

TOBLER, W. R. (1970) *Selected Computer Programs*, Michigan Geographical Publications, Dept. of Geography, University of Michigan, Ann Arbor, MI.

12

Cartographic Computer Programming

12.1 LANGUAGES

The control required for the implementation of cartographic data structures, for performing cartographic transformations, and for anything other than applied computer cartography requires the use of a programming language. Many students find their way into analytical cartography from other disciplines, and as such are excluded from a more indepth understanding of analytical cartography because they do not know how to program. While much of the material presented in this book can be understood, and even taught, without the use of programming, the advanced student may wish to go further. If your background has not been in computing, the best advice is to learn programming from the experts. This usually means in a computer science department. Learning programming along with a graphics course is not to be recommended.

When we program we usually do so in a programming language. Just as we use the English language to express our ideas to others, so we use a programming language to express our ideas to the computer. Computer programming languages perform sequences of instructions. This is the case in spite of the fact that our thoughts are often not sequential. A program moves from step A, to step B, to step C, and the move to step C does not start until B is reached. Human thought can easily branch from A to C, or to X, Y, and Z, or all four simultaneously, based on the loosest of connections.

Usually, a program or set of programs written in a programming language performs a specific task. Such a program is referred to as an applications program. The other general division of programming is systems programming, in which the programs control the operating system itself. All cartographic programming falls into the applications programming category, although there is much that cartographers have to gain from an understanding of systems programming.

All computer programs perform certain generic processes. Most programs perform *input*, that is, they read data from files or a user, they perform *operations* which transform the data in some way, such as between data structures, and then they produce *output*. The output can be a column of numbers, a new file, a graphic, or it can be a map. The most important function of a computer programming language is to control the operation that is taking place. The programming language gives complete control over an application; it allows us to say exactly what we want to do, how we want it to be done, what data should be processed, and where we want to put the output.

Languages have different levels. At the very lowest level of instruction computers use programs called microcode. Microcode is in binary and is so fundamental to the operations of the computer that without it you cannot even load the operating system or use the computer. Microcode instructions tell a computer where to find the operating system, what kind of computer it is, and how to start the boot or initialization process.

At the next level is the machine language program. This language has to be tailored exclusively for a particular brand of computer, a particular type of memory, etc. Programs in machine code (language) talk directly to the hardware. Machine code has the disadvantage that not only are data required to be in binary but the instructions are, too. The last of the low-level computer languages is assembly language. Assembler is compiled in the same way as higher-level languages. This means that a program can be entered and stored in a file, then processed into machine language via a large translation program called a compiler. The machine language program is then ready to run.

An assembler gives access to *registers* or parts of the memory into which we can write instructions and numbers. Assembly language uses as building blocks an instruction set, consisting of multi-character mnemonics standing for individual operations. These operations codes perform simple operations such as putting a number into memory, or taking the current number that is in one register, and performing a binary AND with the number in the current register. Assembly language programs are highly adapted to the architecture of the computer on which they run, and as such are capable of running very fast. In many cases, programs are written in higher-level languages, and their most time-consuming modules are rewritten in assembly language for efficiency.

The remaining levels of computer programming languages fall into what are called high-level languages. High-level languages have named operations, sequences of characters which perform control, and defined data types. An example of a widely distributed higher-level language is FORTRAN, or FORmula TRANslation language, although it certainly is not the only one, and BASIC, COBOL, and others found their way into general use. A distinguishing characteristic of these languages is that although they contain the mechanics to write modular programs, they do not make the modular structure an integral part of the language.

One step beyond general high-level languages like FORTRAN are the structured languages. Structured languages, like Pascal and Ada, encourage the user to write programs that are modular. Modular programs consist of independent units that can be worked on in isolation from each other. A structured language encourages the splitting up of programs into smaller and smaller pieces. Each small program can be worked on, tested, and optimized independently, and then the modules can be assembled into a whole, working program. Other structured languages are PL/1, ALGOL, APL, C, and RATFOR. In addition, a number of special purpose languages exist at this level, which are suited to specific tasks, such as text and list processing, and complex data-base management. Among these are LISP, PROLOG, MODULA, SNOBOL, and the object-oriented languages such as Smalltalk. Beyond the high-level language are application-level programs or macros within systems with their own internal programming language. Graphics languages exist which can perform manipulation of graphic objects and data structures with a command language rather than a programming language. For example, within AUTOCAD, you can program with LISP, or in many systems you can query the data-base using SQL (Standard Query Language).

The language used throughout this book is a language called C. C is the third version of a programming language developed at Bell Laboratories by Dennis Richie. Originally written on a DEC PDP-11 for the development of the UNIX operating system, the language is general purpose, terse, supports advanced control and data structures, and is free of many of the restrictions placed on other languages by their operating environments. The language was chosen for this book because C language programs are brief, powerful, support the complex data structures required for cartography, and because C is highly portable to different computers and operating systems. The language was originally set forth in a book, *The C Programming Language* by Brian Kernighan and Dennis Richie (1978). This book set the early C standard (K and R C), and is the standard followed in this book. Since then, the American National Standards Institute has endorsed a version of C (ANSI-C). The second edition of Kernighan and Richie's book (1988) contains both standards and highlights the differences between them, which to an experienced programmer are minor and are almost entirely limited to type definitions.

The C language gains in brevity by leaving out all special purpose functions, such as mathematical operators, input/output, and text handling, and instead making it easy to build and use program libraries containing working algorithms for these functions. Thus, for example, where FORTRAN defines a SQRT operator to compute a square root, C instead assumes that the user will link the program with a "mathematical library" containing an implementation of an algorithm for computing the square root of a number. C compilers as a result can be relatively small, and many different compilers, debuggers, and programming aids are available for C, even on microcomputers.

12.2 GOOD COMPUTER PROGRAMS

Given a group of programmers with different levels of experience and a problem, perhaps even a problem solved by a published algorithm, it is most likely that each programmer will write a different program to solve the problem. Some programs will not work, in which case it is easy to tell whether or not the solutions have value. Among the working programs, however, how can we tell which is the best, or even which are good? To be able to do so requires that we can tell a good program when we see one.

Fortunately, good computer programs have some distinguishing characteristics. So much so, in fact, that it is possible to have a good program that doesn't work and a bad program that does work! In this section, we will try to cover some of the factors which make computer programs "good."

To begin with, a good computer program is *readable*. This means that programmers can read the program, working their way through it without spending an undue amount of time trying to figure out what the original programmer meant. "Clean code" is neatly formatted (usually by a "beautifier program") and is well presented. A good computer program is *structured*. A structured program has embedded within itself subprograms which do smaller and smaller tasks, so the main program consists of very general tasks. This program passes program control to more and more specific tasks, each of which works independently of the others, and returns control when the subtask is complete. This "task nesting" can make programs very manageable and can reduce programming time significantly. Structuring within some languages such as within FORTRAN is by function and subroutine. In the C language subroutines do not exist, only functions. So in C a program is defined as a function which is itself a set of functions.

The philosophy behind structuring is that programming complexity is worth minimizing. Writing tiny programs that do very specific tasks is comparatively easy, so if we can take a major task and divide it down into many small tasks, each of which can be solved by writing a very simple program, programming complexity is minimized. This approach could be called the "divide and conquer" approach because a difficult problem, divided into less difficult problems, and finally into easy problems, expresses the original problem in a way which can be fairly easily solved. Good programs are structured, and the structure is meaningful in the terms of the problem and works as a solution.

Good programs are *concise*, meaning that they do not contain any more instructions than are necessary to get the task done. A major cause of "verbose" programs is the rewriting performed when program maintenance is done. Rather than reworking the solution, the programmer simply bypasses the old instructions and adds new, duplicating whole sections. Modular programming is by definition concise. When modules are maintained, it is easy to remove a task and rework the solution as a new module or function.

Good programs should be *efficient*. Efficiency is usually measured in terms of the time taken for a task to complete. Many simple measures can be taken to make programs efficient, some of which can make vast improvements in execution time at very little cost. Good programs should be *usable*. Programs which are not usable probably have only one user, the author. Worse than this, once the author is finished using the program it will be forgotten. Modular programs encourage usability because modules can be interchanged. A programmer using structured programming rarely starts a new program from scratch. General purpose functions can be used and reused in a variety of different programs and contexts.

Good programs are *documented* and *maintained*. Documentation, it is arguable, is more important than the program. Documentation can be internal, in the form of headers and comments, or external in the form of written text, either interactive (such as help screens) or paged. External documentation should be able to function in a user or tutorial capacity for the new user, and as reference for the experienced user. Often this means two manuals, a user manual and a reference manual. Good documentation is understandable; it can be read by a novice user and understood.

Good software is *maintained*. Good software has somewhere near the beginning a version number, and the version number is high and decimal, such as 4.03. Constant updates mean that a programmer, preferably the author, has gone to the trouble of finding out what is wrong with the software and has changed it to make it better. Usually, for small changes the decimal place is increased. For major revisions of a program, the version number is increased.

Finally, good programs *work*. This means that they produce the correct answer, which is not always the answer we expect or that we want. Working programs must be reliable in two ways. First, they should survive changes in the operating system. Second, programs should be internally consistent. For example, a statistical program should give the same results using the same data more than once. This implies reliability in terms of consistency and in terms of resilience to change.

We have already noted the special demands of cartographic and geographic data. The large volumes of data which need to be handled influence how we write programs which work with geographic data. Geographic programs also have to run on a variety of different types of computers because there is no one computer that is best for every geographic problem. Cartographic computer programs very often use graphic output rather than sets of statistics or numbers. Cartographic programs use geographic data structures instead of the data structures that have been optimized for general purpose data as in data-base management or numerical computing. Some of the more interesting and challenging data structures that exist are those that are specifically designed to deal with map data. Cartographic programs usually have a very different audience from general purpose programs.

12.3 GRAPHICS, STANDARDS, AND GKS

Graphics are extremely important for analytical and computer cartography. Using graphics with a programming language requires some form of interaction between the program and the graphics. This is accomplished using a language binding. There are really two different types of graphics systems, each with different ways of achieving a binding. Type 1 is extremely device specific, and involves a direct or indirect message to the graphics board, the special purpose computer which controls the graphic display. Many graphics programs use the memory addresses on the graphics card directly and send values to the locations which cause the display to react.

For example, on an IBM PC using EGA graphics in mode 6, memory locations starting at address B800 (hexadecimal) and occupying 640 by 200 binary pixels control the screen display. A binary array written to these locations would automatically appear on the display, in fact very quickly. We are virtually talking directly to the hardware.

Although very fast, programs that are written in this way usually only work on one type of graphics configuration and on one specific computer. Different computers have different screen sizes, different colors, different memory addresses and address ranges, and different ways of mapping a byte onto bits, and support different numbers of colors, line thicknesses, etc. Not only must each graphics program be rewritten for each display device, it must be rewritten for each computer. In a rapidly changing hardware environment, to be device-specific means almost instant obsolescence.

As a response, many software manufacturers offer "graphics toolkits," which allow some degree of program portability, but are still very device specific. Many programming languages, such as microcomputer-based versions of languages like C and Pascal, offer graphics extensions or functions which produce graphics. At the very simplest, a MOVE TO and a DRAW are provided, which allow the user to move to a screen pixel location (the geometry of which is the responsibility of the programmer), with or without drawing a line from the current location. Many toolkits support some sophisticated options such as graphics text, circles, ellipses, and polygon fills, but the programming level is still fairly close to the machine, and the resulting programs, while a little more portable, are still highly device specific.

The alternative is to use graphics standards. Two major graphics standards have found their way into analytical and computer cartography. The first of these, CORE, was an outgrowth of a workshop group of SIGGRAPH, the Association for Computing Machinery Special Interest Group on Graphics in April 1974. The early standard was published in 1977, with completion in 1979. This standard was important in establishing terminology and concepts. The standard was broadly implemented, mostly with FORTRAN bindings, during the 1970s and 1980s. Starting in 1979, an ANSI Technical Committee on Computer Graphics used the CORE system as a model in designing a new two-dimensional standard called GKS. Almost all of the facilities available in CORE were carried over into GKS and its three-dimensional equivalent, GKS-3D.

CORE showed users the advantages of separating modeling and viewing functions. This means that a graphical "object" such as a cartographic object, can be defined, given attributes, etc., independently of the geometry and properties of the final image, an idea termed "virtual maps" (Moellering, 1983). CORE's weaknesses were the concentration on line-drawing functions, its inability to support a multi-user, workstation environment, and the failure to carry the standard over to include language bindings. The latter meant, for example, that function and subroutine names, the order of arguments in function calls, and even the number of arguments and their meanings were changed by individual software producers. This led to a failure of CORE-based programs to be very portable, one of the principal advantages of using a standard in the first place. The CORE system has not been considered for formal adoption as a standard by any ISO or ANSI committee, and will not be in the future since most of its concepts are now embedded in GKS.

The purpose of using a graphics standard is to ensure that a program written using the standard will survive for new systems, be portable, and work independently of the specifics of implementation. Using standards, we sacrifice direct access to a device, although GKS provides a high-level of specialized access through the generalized device primitive and special escape sequences supported by the language binding. The ability to support such a generic work environment is achieved by separating the graphics into the graphics system, the language binding, and device drivers. Device drivers are computer programs which convert graphics operations into instructions that communicate directly with the hardware. This requires a device driver for each particular device we are going to use, so that users must add device drivers as new devices are acquired. For the programmer, the trade-off is flexibility for drawing speed. Device independence also means that the same map can appear on a printer, on an electrostatic plotter, on a camera, or on the screen of the terminal and it will look the same.

As we saw in Chapter 9, with a graphics standard we use three different coordinate systems. We call the coordinate system of the application the world coordinate system. Graphics standards use normalized device coordinates (NDCs), which are coordinates on a virtual device. A virtual device is an idealized planar space on which the map is to be drawn. Under GKS, the lower left-hand corner of the NDC space is at [0,0] and the upper-right hand corner is at [1,1]. In the CORE standard, which supports three dimensions directly, the center of the screen is [0,0,0], the upper right-hand corner is [1,1,0], and the lower left-hand corner is [−1,−1,0]. At the display end, there are workstation or device coordinates which are locations on the display surface as referenced using the mechanics of the display system. Under a graphics standard, the specifics of the device coordinates are unnecessary, although we can inquire their values if need be. In addition, once a window and viewport transformation have been established, the programmer can reference all locations with map coordinates, and specific locations on the map such as locations of legends, titles, insets, etc., in normalized device coordinates.

Graphics standards also allow maps to be structured. At the lowest level, the map consists of primitives, Primitives can be collected into *segments*. Segments can be named, numbered, and recalled by reference instead of being rebuilt from data. Different transformations can be applied to whole segments. Similarly, higher-level implementations of the standards support the *metafile*. The metafile is a generic storage space which behaves the same as a device or workstation. Thus to save the outline of an island, a collection of polylines, they could be named as a segment, written to a metafile, and then later during the same program run, or even later by another computer program, retrieved and redrawn with different transformations or attributes. It is sometimes possible to retrieve these metafiles from different programs, perhaps even in different languages.

An additional feature of the standards is the ability to cluster groups of attributes into *bundles*. A bundle can be selected and uniformly applied to either segments or primitives. For example, we could define a set of attributes which should apply to map titles, such as bold-face fonts, large letters, a specific spacing of letters, a particular color, and text centering. By referencing this bundle, we can apply it to a specific piece of text which is to be a title, rather than changing all the attributes as we go.

A typical computer graphics program sets the attributes after having computed where it wants to draw primitives, opens a segment, simultaneously defines and draws a segment, and then closes the segment. For example, to draw a solid red polygon first we need the boundaries of the polygon, usually from data. Next we assign characteristics, such as fill it with solid red. With this done, we simply open a segment, draw the polygon, and finish. Once drawn and defined, the red polygon can be recalled very easily just by referring to its segment identifier.

In Chapter 13, we will see how to write a graphics program using the GKS standard. The GKS standard has become an important part of graphics programming, especially since the standard's adoption by the American Standards Institute and by the International Standards Organization in 1985. The standard includes a functional description of GKS (ISO 7942:1985), the metafile (ISO 8632-1,2,3,4:1987), and bindings for FORTRAN (ISO 8651-1:1988), Pascal (ISO 8651-2:1988), and Ada (ISO 8651-3:1988). The three-dimensional extension of the standard was approved by ISO in 1988 (ISO 8805:1988). The ISO review of the standard involved over 100 scientists and about a 50 man-year labor effort. The standard has been adopted for several books in computer graphics, including Enderle, Kansy, and Pfaff (1984) and Hearn and Baker (1986). While other standards, discussed in the following section, are often more suitable for specific graphics and even some cartographic applications, GKS remains the benchmark against which all computer cartographic software in the future will be measured.

12.4 HIGHER-LEVEL GRAPHICS LANGUAGES AND STANDARDS

Standards in the computer graphics industry have been slow to develop, yet offer a large number of advantages to the programmer. A standard expresses a nominal set of requirements, a minimal level of performance, and a mandatory area of compliance and conformance to accepted specifications (NCGA, 1987). In the United States, it is the American National Standards Institute (ANSI) which reviews and approves standards and represents the United States in the International Standards Organization. It is the X3H3 committee of ANSI which started examining standards for computer graphics in 1979. The X3H3 committee initiated GKS, the Computer Graphics Metafile (CGM), the Programmer's Hierarchical Interactive Graphics Standard (PHIGS), the Computer Graphics Interface (CGI), as well as GKS-3D.

GKS-3D extended GKS to handle the definition and viewing of three-dimensional "wire-frame" objects. The system supports hidden-line and hidden-surface removal in the workstation transformation, but does not support rendering techniques such as shading and light sources. GKS-3D places a significantly enlarged set of demands upon the host computer, particularly the amount of memory used and the processing power required. As such, microcomputer applications may be limited to only the most powerful machines.

The Computer Graphics Metafile is a file format suitable for saving a "snapshot" of graphics in their final form. Access to images stored in a metafile is either sequential or random access, and the image is defined in a device-independent way. The purpose of the CGM is to allow images generated on different graphics systems to be interchanged and regenerated on different computing and display environments. The three parts of CGM contain the character encoding for assisting in network communications, binary encoding to support transfer between machines with different word sizes and operating systems, and clear-text encoding to ensure maximum readability for use and debugging.

The Programmer's Hierarchical Interactive Graphics Standard, like GKS, is a functional definition of the links between a programming language and a graphics system. PHIGS supports some operations unavailable in GKS, including the ability to nest graphical objects hierarchically (important in modeling applications and CADD), and graphical object data-base support. PHIGS is especially useful when complex objects are to be constructed from smaller simpler objects; for example, a fan can be constructed by defining a blade and then repeating it with rotation. PHIGS supports three dimensions and interactive graphics, such as panning and zooming, at the workstation level. PHIGS increase in flexibility and complexity, however, means that many workstations, particularly at the lower end of the power scale, are not capable of supporting all of the capabilities of a full PHIGS implementation. PHIGS has come to be used, therefore, in high-end interactive workstation applications, particularly in computer-aided design (CAD), architecture, and computer assisted manufacturing (CAM).

The Computer Graphics Interface (CGI) specifies a set of basic elements for the control of data exchange between the device-independent and device-dependent levels in a graphics system. This involves a standardized virtual device interface (VDI). CGI can operate as software to software, in which case it transfers graphics from metafile to binding, or vice versa. As a software-to-hardware link, CGI allows stored graphics to be output to any supported device independently of the graphics system which generated them. Thus a map stored in CGI could be reproduced using the available display devices, or retrieved for analysis via a computer programming link. The most important function of CGI is to allow device-specific graphics systems to produce output which is in accordance with standards, and so to support a larger number of devices.

A large number of additional graphics systems exist at this higher level. For window-based applications, the X-windows system from M.I.T. has gained widespread acceptance. In addition, standards for graphical rendering, and even entirely open graphics-based operating systems, are now in the works. It is important to remember that to be of use a standard should be both supported and used. Programs written to any standard will not have to be rewritten as hardware and software change flavor with the times. If the early days of computing can be classified as the era of hardware, then the next generation will clearly be the era of software, and above all, programming standards. Links with the standard version of C, now approved at the ANSI level, and with the Digital Cartographic Data Standards, ensure that the duplication of effort and time wasting of the early days of analytical and computer cartography are over, and that cartographers can concentrate upon the more substantive aspects of their discipline, safe in their knowledge of how to produce the map.

12.5 REFERENCES

ENDERLE, G., KANSY, K., AND PFAFF, G. (1984) *Computer Graphics Programming: GKS—The Graphics Standard*, Symbolic Computation, Springer-Verlag, New York.

HEARN, D., AND BAKER, M. P. (1986) *Computer Graphics*, Prentice-Hall, Englewood Cliffs, NJ.

International Standards Organization (1985) *Information Processing Systems—Computer Graphics—Graphical Kernel System (GKS) Functional Description*, ISO Publication Number 7942:1985.

KERNIGHAN, B. W., AND RICHIE, D. M. (1978) *The C Programming Language*, Prentice-Hall Software Series, Prentice-Hall, Englewood Cliffs, NJ.

KERNIGHAN, B. W. AND RICHIE, D. M. (1988) *The C Programming Language*, Prentice-Hall Software Series, Prentice-Hall, Englewood Cliffs, NJ, Second Edition.

MOELLERING, H. (1983) "Designing interactive cartographic systems using the concepts of real and virtual maps," *Proceedings, AUTOCARTO 6*, Sixth International Symposium on Computer-Assisted Cartography, Ottawa, Ontario, October 16-21, vol. 2, pp. 53-64.

National Computer Graphics Association (1987) *Standards in the Computer Graphics Industry*, National Computer Graphics Association, Fairfax, VA.

13

Writing Cartographic Software

13.1 THE PROGRAMMING ENVIRONMENT

Writing a computer cartographic program rather than simply using an existing program establishes the degree of control over producing maps necessary for analytical cartography. All too often, however, the decision to write a program is taken too lightly. The result is often confusion, discouragement, and worst of all, a waste of time and resources. With a little thought and planning, however, computer programs which produce maps and map-based analyses can be made to work, and to work well, without the hours of grueling work now known as "hacking."

This development has come about because of some significant improvements in the methods of computer programming. While programming has been around for many years, it is only recently that "software engineering," the application of engineering and computer principles to the act of programming, has turned the "art" of computer programming into more of a science. Software engineering considers every aspect of a piece of software, from its purpose to its maintenance, to the personnel who will produce it, in addition to the language, coding, and debugging of the program.

The first stage of writing a computer cartographic program is the design. During this stage, we should ask what the software will accomplish, what will be expected of the user, how the software will be documented, maintained, and used, and what its expected lifetime is. Professional software engineers accomplish this step with group meetings and consider a large number of possible alternatives before making selections. Once the software's purpose, documentation, and user interactions are decided upon, the next round of decisions should involve a schedule, the choice of a language, a strategy for debugging, checking and verification, and a plan for the software's use and update. The schedule is important. Programming can lead to large amounts of time being wasted on incorrect solutions, and often the timetable is the only force directing the programmer to give up and try another way. The first draft of a user manual and reference guide, or the writing of on-line help facilities, is a good way to help in planning, organizing, and making design decisions relating to the software. Typically, during the early phase of the design, the software is seen as being able to accomplish too much, so a realistic trimming down is often necessary.

In the next phase of design, the software itself is started, but no programs are written. The components are broken down into modules, and decisions are made about how interaction between modules is accomplished. Modules should be chosen so that they are logical divisions of the program, which can be programmed, tested, and maintained in isolation from the program as a whole. At this stage, the places in the program where interaction with the user is necessary must be determined, and the prompts the user is to receive should be planned. Decisions are made about the valid responses to prompts, and system defaults which will be taken without the intervention of the user are planned. Finally, the programming language is chosen, and planning begins for the use of the *programming environment.*

The programming environment is the entire set of tools available to the programmer to produce working and correct programs. An ideal environment should include an editor, a debugger, a syntax checker, a beautifier, a compiler, and a linker. The *editor* is the primary tool by which the programmer enters control statements, data, and documentation into the computer. Most people have their favorite editor, and a decision to stay with one's favorite is sound, since many editors have now been converted for use in a large number of programming environments. As a minimum, the editor should be highly interactive, should allow the partial and complete reading and writing of files, file concatenation (joining files together), searching for text strings, simple movement through the file, global searching, and substitution (making changes to the whole file), and should be easy to learn and remember. While allegiance to one editor is one way of remembering commands, the additional features offered by an alternative editor may well be worth a few hours spent reading manuals and running through learning exercises. With the editor, the programmer can enter into a file the C language computer instructions which produce a map, known as the *source code.*

A *beautifier* is a utility program which reads the source code for a computer program and rewrites it with correct indentation, consistent spacing, and with uniform typing specifications. If you have never used a beautifier, it is difficult to appreciate the fact that correct order within a program can be a very effective debugging aid and even a check for logical consistency. It is not necessary to make programs terse and cramped. Any savings generated by such code can be eliminated by the encoding of a single error which becomes hidden by the sloppy program layout. Many beautifiers are also syntax checkers. A *syntax checker* is a utility program which checks source code files line by line and detects errors in language syntax, that is, inconsistencies in the rules and requirements of the programming language. A good syntax checker will find errors, point the programmer to the exact place where they were found, and give at least a hint as to which rule has been violated. For example, a common C language syntax error is to forget to terminate a statement with a semi-colon. A suggestion by the checker such as "Syntax error line 23" is vastly inferior to "Missing terminating ';' for line 23." It should be noted that using the compiler to find syntax errors is a waste of time, and can lead to a whole new kind of compound logic-syntax error.

The next stage is actually to *compile* the syntactically correct program. With prior syntax error-checking, the compiler is often satisfied fairly easily. This is absolutely no guarantee, however, that the program is correct or even workable. Using C, the final pass of the compiler, called *linking*, has great importance, since C shares program modules between programs. Linking to system libraries such as the math and windows libraries is done at this stage, as also is linking to the libraries which constitute GKS and the GKS-C language binding. Linker error messages are usually to communicate the fact that a named or required library was not found, or its name was misspelled. After a successful compilation and linking, C usually produces two new versions of the program. The first, from the compiler, is the object module, that is, the compiled machine language version of the program or module. After successful linking, another version, the executable program, is produced. Under MS-DOS, typically a program called prog.c will produce an object module prog.obj and an executable program prog.exe. To run the program, the latter is used.

Often, after all of these stages have been completed, the program either produces incorrect results or "crashes." Programs which crash can be started under the control of a *debugger*. Debuggers allow the program to be run, stopped, and the contents of variables listed to determine which variables had what values when, or to see which modules were the source of the error. The sequence *editor* to *beautifier* to *syntax checker* to *compiler* to *linker* to *debugger* may be repeated many times before a program even runs, let alone works and gives correct results (Figure 13.01). Graphics introduces yet another level of debugging. Graphic debugging is assisted by the GKS-C language binding and by the error control built into the GKS standards, which can echo function names and the values passed to them on an ongoing basis. The actual graphics created are also useful debugging tools. Blank screens often mean an error in the transformational geometry involved, while errors with primitives and attributes can generate graphics better suited to the walls of the Museum of Modern Art than to cartography.

Once a program has been written and debugged, it is time to return to the documentation and program purpose to check and verify the program. First, the program should be tested with a small test data set for which the results are known. Few such data sets exist in cartography, so checking with hand calculators is often in order. Attention should be given to special case exceptions to computations such as taking the square root of negative values, or testing for the intersection of vertical, parallel lines. Once small data sets check out, a real-world cartographic data set should be used. Special cases should be deliberately introduced into the data to check circumstances which may not have been foreseen during design. Real data has a habit of introducing these

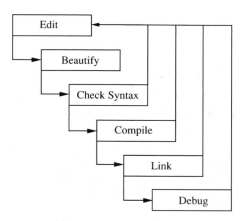

Figure 13.01   The Programming Sequence

special cases, as well as other limitations such as size restrictions, into the program and testing the program's design limitations. Failure of the program at any of these stages may demand going back to a module to make corrections. It may even be necessary to redesign a whole module, which is possible only with structured programs. If, at this stage, it is found that the entire program structure is poor, this is a good time to give up, since no number of program "patches" can repair a poor overall structure or logic. As a final stage, many programs are optimized. This involves analyzing, often using other computer programs, each module to see how much time is spent in each function and at each stage in each function. The most time-consuming tasks are reworked first, each optimization making the program run faster or use less memory. Some C compilers and operating environments, such as UNIX, have built-in optimizers and program run-time analyzers (the -O compiler option and the profile program). In other cases, optimization is a painstaking process, with each improvement taking seconds off the run times for the program.

Having given a description of the programming environment, it goes without saying that it is critical to cartographic software. Much time will be wasted if the programmer avoids using the tools available, tools which are now parts of most mainframe and minicomputer C programming environments, and even most microcomputer programming environments. A little time invested in learning the sequence used in Figure 13.01 can save hours of wasted time and result in much more effective and better constructed computer cartographic software. The programmer should think first and foremost about the map being produced, then about the specifics of the programming language in use. Above all, the design should be organized and planned before a single line of computer code is written. Writing computer programs and designing them are two entirely different tasks, and the worst place to think about design is inside three levels of open control statements.

13.2 THE GKS C LANGUAGE BINDING

In some earlier chapters, particularly in the discussion of the symbolization transformation, the GKS standard was presented as a means by which cartographers can write computer programs which produce maps. We saw that the standard can be implemented at different levels, with different device support and functionality for different levels. We also saw that the link between programming languages and graphics under the GKS standard is the language binding. The ISO specification of GKS lists standard bindings for GKS for FORTRAN, Ada and Pascal. Since the C programming language has only recently been revised into an ANSI version, there is as yet no standard C language binding to GKS. This does not mean, however, that none exist, merely that there are some differences in bindings between the products of different software vendors. Fortunately, the standard is quite general when it comes to the major source of differences, i.e., the naming of functions. Function names are listed by English language names rather than function names. Figuring out a function name for a different binding, or even transporting a piece of mapping software between computers or programming languages, often consists therefore of translating the names of functions, and assuring that arguments in the lists passed to the functions are consistent. A convention used by some vendors for C functions is to use the full rather than abbreviated function names, replacing blanks with underscores (_). On some computers, especially microcomputers, there is a need to abbreviate due to DOS name limitations, or limitations on the linking capabilities of C compilers. In this discussion, the simplest form of the name will be given.

A C language binding can be thought of as a set of predefined C libraries which contain all of the functions necessary to use the standard. The functions break down into those which manage primitives and attributes, those which handle interaction with the user, those which control the workstation, those which perform transformations, those which manage segments and the metafile, those which manage errors, and those which allow the programmer to query GKS to establish any of the above. In all, this makes several hundred functions, since for every primitive, attribute, and interaction function there is a corresponding query function. For the function set_fill_area_color_index(), to give one example, there is a corresponding GKS inquiry function named inquire_fill_area_color_index().

There is a formal structure for the use of GKS functions. The sequence applies for all GKS levels and implementations. The first GKS operation that a program must perform is to open GKS. This performs an initialization of the software on this system for this application. The function used is *open_gks()*. This function sometimes uses as a parameter the name of a file which GKS will use to write GKS error messages. The file should be opened in write mode by the user's program, and the value passed should be a pointer to type FILE. The next stage is to open the workstation. This is accomplished using the function *open_workstation()*. Arguments should include an integer identifier for the workstation to be used in successive workstation calls, and a unique number which identifies the type of workstation to GKS. This number is provided with the GKS language binding, and sometimes with the specific device driver installed.

The number must identify not only the "type" of device, but often the specific model number or screen size. Once opened, the workstation should be activated, or made ready for interaction, using the function *activate_workstation()*. This function simply uses the workstation identifier given to *open_workstation()*.

The "digital canvas" is now ready. The programmer can establish the normalization transformation(s), read any data required, and set any attributes, and is then ready to draw. If no segments need to be defined, the primitives are simply included in the order used. To draw and save primitives in a segment, the function call *open_segment()* is used to create a new ségment, which includes all primitives and their attributes until the *close_segment()* function is used. Once defined, segments can be redrawn, renamed, copied between workstations, or written into the metafile. For example, to redraw the entire map, the function call *redraw_all_segments()* could be used. The way back out of the GKS system is the reverse of the way in. After all segments have been closed, the workstation can be cleared using *clear_workstation()*, deactivated using *deactivate_workstation()*, closed with *close_workstation()*, and finally GKS can be shut down using *close_gks()*. The GKS system also provides an immediate error exit from gks so that the programmer can escape from errors without involving the graphics, *emergency_close_gks()*. The entire sequence is illustrated in the following program segment.

Function 13.01

```
/* ===============================================
/*  Illustrate typical GKS-C language binding
/*  by showing gks startup and shutdown
/* ===============================================
*/
    #include "header.h"  /* Header file to link all necessary
                            gks and other variable & functions */
    #define WS_TYPE 5301 /* Unique number for SUN 3/50
                            workstation */
    main() {
    FILE *error_file;
    int ws_id = 1; /* Numerical identifier for workstation */
    int ws_type = WS_TYPE; /* Numerical identifier for
                            workstation */
    /* First, open up a file for GKS to write error messages */
    error_file = fopen ("gkserrors","w");
    /* Initialize GKS */
    open_gks (error_file);
    /* Open the workstation */
    open_workstation (ws_id, ws_type);
    /* Activate the workstation */
    activate_workstation (ws_id);
    <Generate and draw the map, do any necessary user interaction>
    /* Deactivate the workstation */
```

```
deactivate_workstation(ws_id);
/* Close the workstation */
close_workstation(ws_id);
/* Close down gks */
close_gks ();
}
```

Few differences exist between bindings, with one exception. Many of the C language bindings use single-variable parameters as function arguments. This method follows closely the FORTRAN standard. A line, for example, could be sent to a line drawing (polyline) function using *polyline (number_of_points, &x[0], &y[0])*; This provides the function with the number of points in the vectors x and y, and the starting addresses of the points. Alternatively, at least one major GKS binding uses a structure of type POINT, and instead passes the address of the entire structure to the GKS function.

Each GKS C binding implementation has its own details. Usually, there is a large header file, often called gks.h, which contains all the values of the function arguments in a definition header file, so that, for example, color number 6 on a specific device is predefined as #define RED 6 Also, different implementations split the libraries up differently. Sometimes the language binding is separate from the main gks library. This must be taken into account when compiling and linking C programs.

New GKS implementations, language bindings, and device drivers become available frequently. Trade journals, the so-called "glossies," and graphics conventions are good sources of information about these changes. In addition, several GKS implementations are now appearing as shareware. While these versions are unlikely to support the more exotic computers and graphics devices, the more common ones may be found. In addition, several companies now make add-on software packages built over the GKS. In a few cases, these systems support advanced mapping functions. Other GKS implementations are very low level (0b, for example), do not support many interactive functions, and allow output only to the metafile. These versions come with metafile translators, which convert the metafiles into the device-specific graphics calls of a great number of different graphics devices. Among the most popular are translations to PostScript for output on laser-jet printers, or to protocols of desktop publishing packages, allowing the output from computer programs to be pasted directly into a paper, article, set of documentation, or even a textbook. Appendix B contains the names of some of the companies which supply GKS software.

13.3 WRITING YOUR OWN MAPPING PROGRAM

Assuming that you have read much of this book, have read about the C programming language, understand cartographic data structures, and have gone through the software design process described above, how do you actually go about writing a computer program to generate a map? During the design process, you have asked yourself what kind of map you wish to produce. You have blocked out a set of modules, which can do any necessary input, structure the data, perform any necessary transformations, and then generate the map. The next and most important step is to write the program.

First, you should carefully research the various C compilers available to you. The newsstand microcomputer journals contain large amounts of information to assist in deciding which compiler to buy. If you have access to a larger computer, see if the C language is available and how it is supported. If the computer runs the UNIX operating system, then many of the tools available as part of the programming environment discussed above are available. UNIX can also run on many microcomputers and is the most common workstation operating environment. Do not start writing programs until you are familiar with the language. When you do, make lots of deliberate mistakes, and see how the various parts of the programming environment react to them. If C is your first language, then as mentioned in Chapter 12, you should seek out instruction from a computer scientist. If you are coming to C from Pascal or a similar language, you should have very few problems, but should find an appropriate C language tutorial and work carefully through it. Pascal programmers should not be tempted to rewrite C to look like Pascal (although it is possible). While many of the concepts from languages like FORTRAN will help in the conversion, it is better to learn the structured approach from scratch.

One of the features which eventually becomes the source of many errors in C programs is the distinction between the types of variables. The C language supports integers (unsigned, short, regular, and long), characters, floats, and doubles, and while variables must be assigned types exclusively, type conversions are not automatic. For example, to make a floating point (real) equivalent of the integer number 5, it is not enough simply to assign the value. The type should be explicitly converted in the statement. This is illustrated in the following program segment.

Function 13.02

```
/* =======================================
/*   Remember to explicitly perform type
/*   conversions in C, otherwise beware!
/* =======================================
*/
#include <stdio.h>
    main() {
        int an_integer = 5;
        ofloat a_float;
        printf("To begin, the integer value is %d\n",an_integer);
        a_float = an_integer / 2;
        printf("Halved without type conversion the value is %f\n",
            a_float);
        a_float = (float) an_integer / (float) 2;
        printf("Halved with correct type conversion the value is
            %f\n",a_float);
}
```

One of the best ways to learn about computer cartographic programming is to study the programs written by other cartographers. While the list of available sources is short, it is growing steadily. Cartographers have been slow to publish their computer programs in a form suitable for use by others. As a result, many cartographers have duplicated the work of others without even having the ability to compare the two results. The programs presented in this book are not complete programs in themselves, but are building blocks around which more useful computer programs may be constructed. In most cases, writing a computer program using these functions will involve reformatting some existing digital cartographic data, entering the data into a programming data structure appropriate to the task, performing any necessary cartographic transformations, and producing a map. The cartographic transformations may involve implementing an algorithm, perhaps one published with a descriptive paper or journal article.

If the programming is done effectively, each of these tasks can be performed by one or more program modules. Remember that these modules should become your own personal library of cartographic functions, and that, for best effect, they should be shared freely with your friends and any other interested persons. The best way to find weaknesses and errors and to optimize the code is to release programs to other users. Also, however, remember that as a user of computer cartographic software you have an obligation to report problems, errors, and weaknesses to the author of any piece of software, for without this feedback, no improvement will take place. It is also entirely appropriate to report more favorable items of news!

13.4 CARTOGRAPHIC SOFTWARE AND THE USER INTERFACE

The eventual success or failure of a piece of cartographic software is determined not by innovation, sophistication, or accuracy, but instead by usability. Many excellently written and designed computer programs have been passed over entirely because they lack the elements which make the capabilities of the program accessible to the user. The usability of most software, and this includes computer cartographic software, is therefore determined by the user interface of the program. The user interface is the entire set of means by which the user communicates with software to achieve a specific mapping purpose.

In the early days of computer cartography, most software which produced maps ran in batch mode. The means of interaction with the program was by supplying a set of parameters in a file (or even as punched cards), which were read and acted upon by the software. Information could include the names of files, the number of data points in a file, the precise format of an input record, or the numbers of options listed in the documentation. While batch processing was capable of generating maps, errors were common, since the map output came at the end of a long sequence of operations which happened quite slowly, often overnight. Even generating a map could take several days, but debugging the software for mapping could take weeks or months.

As interactive computer systems became commonplace, many computer mapping systems moved to interactive processing. The earliest user interface in such an environment was the command-driven program. A command-driven program reads one command from the user at a time, and acts upon it either immediately or when a general "action" command is given. Commands usually involve a keyword, such as PLOT or READ, and are followed by a list of values which supply parameters to the program. For example, to read a grid data file, the command

```
READ "Input_file" 200 300 I
```

could be used to read a grid of integers from file *Input_file* with 200 rows and 300 columns. Note that these commands can be placed together into a file and submitted together in batch mode, implying that command mode contains a batch mode.

The next level of sophistication in terms of the user interface is to provide a menu in which commands can be selected by number or letter. Figure 13.02 shows a typical menu for a computer mapping program.

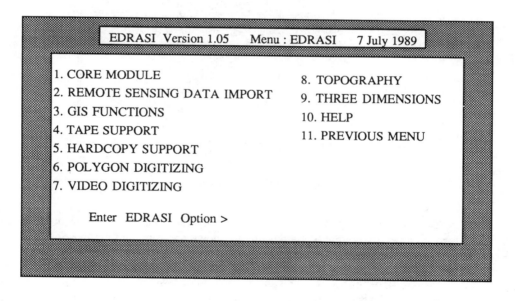

Figure 13.02   A Menu-Driven Software Package

Using a menu, selections can be chosen by entering a number instead of the name of the option, in the same way that you can order from a Chinese menu by asking simply for "number 6" instead of La Zi Ji Ding. Many menu systems also allow the name of the command to be typed, thus supporting command lines also.

Increasingly, operating systems are taking advantage of the ability of computer programs to use windows. Windows are areas of the screen where interaction can take place, or where graphics can be generated. A windowing system allows the windows to be moved, resized, or selected to be visible, often with the use of a mouse (Figure 13.03).

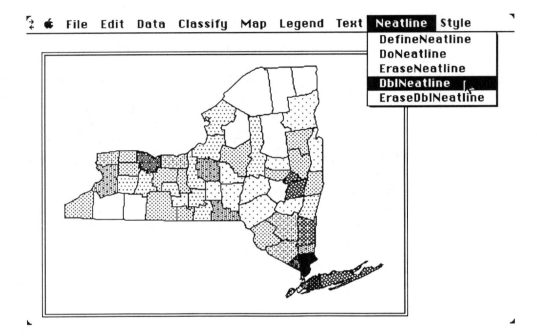

Figure 13.03   A Windows-Based Software Package

Selections can be made by moving to a header and pulling-down the menu, or the next menu level, by holding the mouse button down and moving the mouse. Menus can also appear on the screen at appropriate time during the execution of the program, the so-called pop-up menu. The windows environment has been popularized by microcomputers, especially the Apple Macintosh. Efforts are now under way to specify standard window environments for large and small computers alike, the results of which may be a standard graphic interface with most computers in use. Since the window functions are callable from computer programs, this gives the programmer the ability to use programs which incorporate a consistent interface which is easy for all users, since it embodies concepts which are common between computers.

For computer cartographic programs, it is important that the means of interaction with the user be kept in mind at all times. Modular and flexible programs make it easy to use one command system during software development, and another when the program is finished. A better approach, however, is to write the user interface first. The entire user interaction system can then be debugged, tested, and refined to make the program more "friendly" before the program even works. For this purpose, many large programs are first written with "function stubs," one- or two-line functions which simply report on the fact that they were called in the right place and with the correct arguments. Producing a "skeleton" program in this way makes working on a large program much easier and also allows several people to work on a program at the same time.

The user interface is the last challenge facing the analytical and computer cartographer. As we have seen, the cartographer of today must have an understanding of the hardware and software tools of computer cartography. They must understand cartographic data, cartographic data structures, and the programming mechanisms by which they can be manipulated. The analytical cartographer must understand cartographic transformations and be able to use the power of computer programming and the programming environment to achieve a higher understanding of maps and mapping.

It is the user interface, however, which in the future will determine how computer mapping systems relate to the rest of the world. If cartography is to move away from being a discipline in which a practitioner disappeared for a time and then delivered a map, if cartography is to become democratic enough to place the production of maps into the hands of the non-cartographer, then we must begin by writing mapping software which is usable. This means effective user interfaces and meaningful assistance with decision-making, and it means that much well-written software remains to be produced. The opportunities presented by the wealth of digital cartographic data now becoming available are extraordinary. How effective this revolution in cartography will be in solving the problems of our world, however, depends upon a new generation of analytical and computer cartographers.

13.5 BIBLIOGRAPHY

ENDERLE, G., KANSY, K., AND PFAFF, G. (1984) *Computer Graphics Programming: GKS—the Graphics Standard*, Symbolic Computation, Springer-Verlag, New York.

FOLEY, J. D., AND VAN DAM, A. (1982) *Fundamentals of Interactive Computer Graphics*, Addison-Wesley, Reading, MA.

HEARN, D., AND BAKER, M. P. (1986) *Computer Graphics*, Prentice-Hall, Englewood Cliffs, NJ.

JOHNSON, N. (1987) *Advanced Graphics in C: Programming and Techniques*, Osborne McGraw-Hill, Berkeley, CA.

KERNIGHAN, B. W., AND RICHIE, D. M. (1988) *The C Programming Language*, Prentice-Hall Software Series, Prentice-Hall, Englewood Cliffs, NJ, Second Edition.

LAMB, D. A. (1988) *Software Engineering: Planning for Change*, Prentice-Hall, Englewood Cliffs, NJ.

PRESS, W. H., FLANNERY, B. P., TEUKOLSKY, S. A., AND VETTERLING, W. T. (1988) *Numerical Recipes in C : The Art of Scientific Computing*, Cambridge University Press, New York.

Appendix A

Vendors of Computer Mapping Software

The following is a list of several computer mapping software vendors. This list is probably not complete, and will become increasingly out-of-date over time. The addresses and phone numbers listed are believed to be correct at time of publication. Note that companies periodically withdraw products, change their names, change their phone numbers and addresses, market new products, and so on. Only products that are computer cartographic in nature have been included. Many GIS and image processing systems also have automated mapping capabilities, but are not listed here.

ALTEK Corporation
12210 Plum Orchard Drive
Silver Spring, MD 20904-7802
Phone: (301) 572-2550
Products: Almap, Microdij
Comments: Digitizer drivers for Altek digitizing tablets. PC compatible. Sophisticated capabilities.

American Small Business Computers, Inc.
118 South Mill Street
Pryor, OK 74361
Phone: (918) 825-4844
Products: Designcad, Designcad 3-D, Prodesign
Comments: Inexpensive PC-based CAD, with some advanced capabilities. See review in *Professional Geographer*, vol. 39, no. 2, 1987, p. 232.

ATC Advanced Technology Center
5711 Slauson Avenue, Suite 238
Culver City, CA 90230
Phone: (213) 568-9119
Products: Grafpak-GKS, Grafpak-Core, Contour-80, Twin Term, P-Cubed
Comments: PC-based and other tools for graphics. GKS implementation includes C. Twin Term is advanced graphics workstation emulator for PCs. P-cubed is a toolbox for image applications.

Autodesk, Inc.
2320 Marinship Way
Sausalito, CA 94965
Phone: (800) 445-5415
Products: Autosketch, Autocad
Comments: Autosketch is an inexpensive PC-based CAD. Autocad is a widespread, sophisticated CAD package for many environments. "De Facto" CAD standard. A "must-see." See reviews in The American Cartographer, vol. 13, no. 3, 1986, p. 269, and vol. 14, no. 4, 1987, p. 367; and *The Professional Geographer*, vol. 41, no. 3, 1989, p. 369.

Bassett Geographic
1103 Rudd Avenue
Auburn, AL 36830
Product: Choromap
Comments: Choropleth mapping package for the AMIGA. Supports digitizing.

Beagle Bros., Inc.
3990 Old Town Avenue
San Diego, CA 92110
Product: Beagle Graphics
Comments: Apple II paint program. See review in *The Professional Geographer*, vol. 41, no. 2, 1989, p. 231.

Bridge Software
P.O. Box 118, New Town Branch
Boston, MA 02258
Phone: (617)527-1585
Product: DATASURF
Comments: PC-based display of 3-D fishnets, with HP plotter support. Also has math-based graphing software.

C and C Engineering
Route 4, Box 3986

Lexington, NC 27292
Phone: (704) 787-4131
Comments: PC-based COGO package with mapping option.

Celeris Inc.
21405 Devonshire Street, Suite 216
Chatsworth, CA 91311
Phone: (818) 709-2181
Product: TacPac
Comments: Tactical display for larger computers using Chromatics workstations. GKS interface, vector and point symbol mapping.

CG Compugraphic Corporation
Type Division
90 Industrial Way
Wilmington, MA 01887
Phone:(800) MAC-TYPE
Product: cgType
Comments: Cartographic-quality text fonts for the Macintosh. Many fonts available.

Chadwyck-Healey Inc.
Electronic Publishing Division
1101 King Street, Suite 180
Alexandria, VA 22314
Products: Mundocart/CD, Supermap
Comments: Worldwide data base from DMAs ONC coverage at 1:1M on CD-ROM. PC plotting, searching, map projections. Supports Dr. Halo and Autocad.

Chautauqua Software
Community Research and Information Systems
P.O. Box 1280
Ripley, NY 14775
Product: Polymaps
Comments: PC-based choropleth mapping. See review in *The Professional Geographer*, vol. 40, no. 4, 1988, p. 467.

Civilsoft
1592 North Batavia Street, Suite 1A
Orange, CA 92667
Products: Cogo-PC Plus, Contour
Comments: Sophisticated COGO, CAD, and contouring package. Autocad and Intergraph interface. Broad device support.

ComGrafix, Inc.
302 South Garden Avenue
Clearwater, FL 34616
Phone: (800) 448-MAPS
Products: MapGrafix, MapStar, NAVplus
Comments: Macintosh-based mapping capabilities of good quality, thematic and surface mapping. MapStar and NAVplus support vehicle navigation.

Computer Associates International, Inc.
One Tech Drive
Andover, MA
Phone: (800) 343-4133
Product: CA-DISSPLA
Comments: Fortran Graphics Library. Engineering option supports contouring, projections, and GKS.

Comwell Systems, Inc.
P.O. Box 41852
Phoenix, AZ 85080
Phone: (602) 869-0412
Product: PC-globe
Comments: Inexpensive PC-based world atlas with query and display capabilities.

D.C.A. Engineering Software, Inc.
P.O. Box 955, Main Street
Henniker, NH 03242
Phone: (603) 428-3199
Products: Coordinate Geometry, DTM
Autocad add-on. Survey data entry, cross sections, grids, contours.

Dynamic Graphics, Inc.
Headquarters Berkeley
2855 Telegraph Avenue, Suite 405
Berkely, CA 94705
Phone: (415) 849-3008
Products: Various
Comments: Advanced software tools for terrain modeling, analysis, mapping, and display. Various software libraries, mostly FORTRAN. Expensive.

Electromap, Inc.
P.O. Box 1153
Fayetteville, AR 72702-1153
Phone: (800) 336-6644
Product: Electromap World Atlas

PC-based world atlas on disk or CD-ROM. IBM EGA/VGA support. Some quite pleasant maps.

Electronic Atlas of Arkansas
University of Arkansas Press
McIlroy House, 201 Ozark
Fayetteville, AR 72701
Phone: (501) 575-3246
Product: Electronic Atlas of Arkansas
Comments: First digital cartographic state atlas, supported by university. Needs PC with EGA. See review in *The American Cartographer*, vol. 16, no. 1, 1989, p. 56.

ESRI System, Inc.
Attn: pcARC/INFO Marketing
380 New York Street
Redlands, CA 92373
Phone: (714) 793-2853
Products: ARC/INFO, pcARC/INFO
Comments: Full GIS with strong mapping and topological capabilities. Raster/vector, overlay, digitizing, editing. Data-base support.

Evolution Computing
437 South 48th Street, Suite 106
Tempe, AZ 85281
Products: EasyCAD, FastCAD
Comments: PC-based CAD system. Review in *The Professional Geographer*, vol. 41, no. 1, 1989, p. 88.

FMS/AC
Dennis Klein & Assoc.
38 Miller Avenue, Suite 11
Mill Valley, CA 94941
Phone: (415) 381-1750
Product: FMS/AC
Comments: Facilities mapping module for Autocad. Some advanced mapping capabilities.

Geocalc Software System Co.
P.O. Box 5308
Philadelphia, PA
Phone: (215) 365-5585
Products: COGO, GenericCADD
Comments: COGO package, with full set of computations and plotting.

Geodetic Technology Co.
1350 East Arapaho Road, Suite 238
Richardson, TX 75081
Phone: (214) 231-9583
Product: Geopac
Comments: HP-calculator-based package for geodesy. Map projections, UTM, NAD 27, NAD 83.

Geographic Data Technology, Inc.
13 Dartmouth College Highway
Lyme, NH 03768-9990
Phone: (603) 795-2183
Products: Dynamap, Safari
Comments: Distributes several mapping packages, including pcARC/INFO, and services. Several mapping and data-base products, including TIGER file handler.

GeoGraphics Corporation
1318 Alms Avenue
Champaign, IL 61820
Phone: (800) 255-2255 Ext. 1223
Products: DigiGraph, Measugraph
Comments: Drivers for GeoGraphics digitizers. Supports Autocad.

GeoSpectra Corporation
P.O. Box 1387, 333 Parkland Plaza
Ann Arbor, MI 48106-1387
Phone: (313) 994-3450
Product: ATOM
Comments: Automated topographic mapping from stereo photographs.

GIMMS Ltd.
30 Keir Street
Edinburgh, Scotland EH3 9EU
Phone: (031) 229-3937
Comments: Micro- and mini-based mapping software. Many devices supported, many thematic mapping capabilities. Good cartographic capabilities.

GNIS
James Taylor & Robert Sechrist
Department of Geography and Regional Planning IUP
Indiana, PA 15705
Products: Geographic Names Information System Retrieval Program, MAPSETS
Comments: Interactive retrieval and plotting from the GNIS data. PC-based. Unique software. MAPSETS are base maps for United States, Europe, and the world with PC and EGA plotting.

Golden Software, Inc.
P.O. Box 281
Golden, CO 80402
Phone: (303) 279-1021, (800) 333-1021
Product: Surfer
Comments: Relatively inexpensive contour and 3-D mapping program. Extensive control, many supported devices, excellent interactive system with on-line help. See reviews in *The American Cartographer*, vol. 13, no. 3, 1986, p. 266, and *The Professional Geographer*, vol. 40, no. 3, 1988, p. 345.

Hasp, Inc.
1411 West Eisenhower Boulevard
Loveland, CO 80537
Phone: (303) 669-7901
Products: Various
Comments: Full survey, COGO, 3-D graphics support. Plotting, digitizing, data-base link.

Image Mapping Systems
P.O. Box 31593
Omaha, NE 68132
Product: MacChoro
Comments: Interactive Choropleth mapping package for the Macintosh. Excellent learning tool. See review in *The American Cartographer*, vol. 14, no. 1, 1987, p. 69.

Instant Recall
P.O. Box 30134
Bethesda, MD 20814
Phone: (301) 530-0898
Product: USA Display
Comments: State data-base system with information retrieval and plotting.

Intergraph Corporation
One Madison Industrial Park
Huntsville, AL 35807
Phone: (205) 772-2000
Products: Various
Comments: Several GIS, CADD, and automated mapping products. Also manufacture workstations.

Lietz
9111 Barton, Box 2934
Overland Park, KS 66201
Phone: (913) 492-4900

Product: Contour Plus
Comments: Contouring, 3-D grids, and COGO. Links to survey and field notebook software.

Mapics Ltd.
26 Bedform Way
London WC1H 0AP, UK
Product: PC-Mapics
Comments: Mapping, digitizing, graphing for the PC. Review in *The Professional Geographer*, vol. 40, no. 4, 1988, p. 466.

MapInfo
200 Broadway
Troy, NY 12180
Phone: (518) 274-8673
Product: MapInfo
Comments: PC-based package with extensive mapping capabilities. Supports data-base, address matching, and point and boundary analysis.

Mapware
P.O. Box 50168
Long Beach, CA 90815
Product: The Map Collection
Comments: Choropleth, symbol, graduated symbol, and surface mapping on the PC. Good educational package. Inexpensive.

Media Cybernetics, Inc.
Imaging and Graphics Technology
8484 Georgia Avenue
Silver Springs, MD 20910
Phone: (301) 495-5964
Products: Image-Pro, Halo, Dr. Halo
Comments: Halo and Image-Pro are software libraries to support graphics and image processing. Dr. Halo is a raster mode paint program. Widely distributed and used software.

Microconsultants
P.O. Box 3455
Mililani Town, HI 96789
Phone: (808) 623-6573 (7-10 pm, Hawaiian Time)
Product: MicroCAM
Comments: PC implementation of the CAM system, with many projections for the world data banks. Excellent instructional tool, extensive control over mapping.

Micromaps
P.O. Box 757
Lambertville, NJ 08530
Product: MacAtlas
Comments: World, U.S., and county outline maps for use with MacDraw and MacPaint.
Review in *The Professional Geographer*, vol. 40, no. 2, 1988, p. 243.

MicroSurvey
Marketing International Inc.
P.O. Box 112
Queensville, Ontario L0G 1R0
Phone: (416) 853-6773
Product: MicroSurvey
Comments: COGO package with plotting and Autocad support.

National Decision Systems
539 Encinitas Boulevard, Box 9007
Encinitas, CA 92024-9007
Phone: (619) 942-7000
Product: PC-MAP
Comments: PC-based data handling and plotting software with thematic mapping capabilities. Review in *The Professional Geographer*, vol. 40, no. 1, 1988, p. 112.

National Planning Data Corporation
P.O. Box 610
Ithaca, NY 14851-0610
Phone: (607) 273-8208
Product: Map Analyst
Comments: Sells marketing analysis and services using own software.

Oedware
P.O. Box 595
Columbia, MO 21045-0595
Phone: (301) 997-9333
Products: HyperDraw, PC-Key-Draw
Comments: Inexpensive CAD for PC with Hypercard-like capabilities.

Pink Software Dev.
Milan Systems America, Inc.
8351 Roswell Road
Atlanta, GA 30350
Product: TurboCAD
Comments: CAD package for PS/2, PC with student discount. Review in *The Professional Geographer*, vol. 41, no. 2, 1989, p. 235.

Plus III Software, Inc.
One Dunwoody Park
Atlanta, GA 30338
Phone: (800) 235-4972
Product: terraCADD
Comments: COGO package with terrain mapping capabilities. Contouring, 3-D.
Link to Autocad.

Precision Visuals, Inc.
6260 Lookout Road
Boulder, CO 80301
Phone: (303) 530-9000
Products: DI-3000, GK-2000, Contouring System
Comments: FORTRAN subroutine libraries for graphics. Core and GKS implementations. Contouring is library for contour and 3-D fishnet display.

Rand McNally & Co.
P.O. Box 7600
Chicago, IL 60680
Product: RANDMAP
Comments: Choropleth and dot density maps from state and other boundary files. PC-based. See review in *Professional Geographer*, 1986, vol. 39, no. 2, p. 233.

RDS Systems, Inc.
6110 Executive Boulevard, Suite 605
Rockville, MD 20852
Phone: (301) 984-1989
Product: MAPS
Comments: PC-based map projection system, includes world data banks.

Rockware, Inc.
7195 West 30th Avenue
Denver, CO 80215
Phone: (303) 238-9113
Products: GRIDZ0, SURVEY, ROCK-SOLID
Comments: Geological software collection, including contouring and 3-D fishnets, COGO, and block-diagram models of subsurface.

SAS Institute Inc.
Box 8000
Cary, NC 27512
Phone: (919) 467-8000
Product: SAS/GRAPH
Comments: Large statistical package with mapping capabilities. See review in *The American Cartographer*, vol. 14, no. 2, 1987, p. 164.

Schroff Development Corporation
4732 Reinhardt
Roeland Park, KS 66205
Phone: (913) 262-2664
Product: Silver Screen
Comments: Micro-based 3-D CAD software with solid modeling capability.

Scientific Software Group
P.O. Box 23041
Washington, DC 20026-3041
Phone: (703) 620-9214
Products: GWN-DTM, GWN-COGO
Comments: Sophisticated terrain modeling. PC-based with Autocad support.

SCO Inc.
740C South Pierce Avenue
Lousville, CO 80027-9989
Phone: (303) 666-7054
Product: GRAFkit (International Computer Exchange)
Comments: FORTRAN GKS and CGM library, plus 3-D plotting and contouring.
Many supported devices.

Select Micro Systems, Inc.
40 Triangle Center, Suite 211
Yorktown Heights, NY 10598
Phone: (914) 245-4670
Product: MapMaker
Comments: Macintosh mapping software with ability to write MacPaint and MacDraw
files. Interactive thematic mapping, with easy-to-use menus. Has own data sets. See
review in *The American Cartographer*, vol. 15, no. 2, 1988, p. 205.

Silicon Beach Software, Inc.
9770 Carroll Center Road, Suite J
P.O. Box 261430
San Diego, CA 92126
Phone: (619) 695-6956
Products: Super 3D, Superpaint
Comments: Macintosh solid modeling and CAD, with animation. Superpaint is
advanced paint package.

Software Concepts, Inc.
45 Church Street
Stamford, CT 06906

Phone: (203) 357-0522
Product: Concepts Computerized Atlas
Comments: PC-based atlas. See reviews in *The American Cartographer*, vol. 14, no. 4, 1987, p. 370, and *The Professional Geographer*, vol. 40, no. 1, 1988, p. 107.

SPSS Inc.
444 North Michigan Avenue
Chicago, IL 60611
Phone: (312) 329-2400
Product: SPSS Graphics
Comments: Large statistical package with mapping capabilities. See reviews in *The American Cartographer*, vol. 14, no. 2, 1987, p. 169; and *The American Cartographer*, vol. 16, no. 1, 1989, p. 57.

Strategic Locations Planning Inc.
4030 Moorpark Avenue, Suite 123
San Jose, CA 95117
Phone: (408) 985-7400
Product: ATLAS*GRAPHICS
Comments: Advanced thematic mapping package for PC-based systems. Supports many devices, includes datasets and base maps. See reviews in *The American Cartographer*, vol. 16, no. 2, 1989, p. 134, and *The Professional Geographer*, vol. 41, no. 3, 1989, p. 367.

Synercom Technology, Inc.
2500 City West Boulevard, Suite 1100
Houston, TX 77042
Phone: (800) 334-7101
Products: Various
Comments: Advanced GIS, automated mapping, and CAD capabilites.

Systat Inc.
1800 Sherman Avenue
Evanston, IL 60201
Phone: (312) 864-5670
Product: Systat
Comments: Statistical package with interesting mapping capabilities. Map projections, Voronoi diagrams, Chernoff faces, contouring, 3-D fishnet. PC, CP/M, Unix, and VMS.

Technical Advisors, Inc.
4455 Fletcher Street
Wayne, MI 48184
Phone: (313) 722-5010

Product: Tech-Mac PC
Comments: PC-based COGO and survey plotting software.

Terra-Mar Resource Information Services, Inc.
1937 Landings Drive, Mountain View, CA 94043
Phone: (415) 964-6900
Products: MicroDLG, TerraPak
Comments: Computer mapping modules for a GIS/Image processing system. Can import/export various data formats.

Uniras, Inc.
50 Mall Road, Suite 206
Burlington, MA 01803
Phone: (617) 272-7260
Products: Unimap, Uniedit, Raspak, UniGKS
Comments: Advanced FORTRAN libraries for graphics, geophysical mapping, raster graphics, contouring, 3-D plotting.

Wild Heerbrugg
24 Link Drive
Rockleigh, NJ 07647
Phone: (201) 767-1100
Products: WildSoft, Wild Cip
Comments: Cross-sections, COGO, CAD, advanced contour interpolation. Strong surveying link.

WORLD
Philip Voxland, Social Sciences Facilities Center
University of Minnesota, MN 55455
Phone: (612) 625-8556
Product: WORLD
Comments: A map projections package based on the world data banks. See reviews in *The American Cartographer*, vol. 16, no. 2, 1989, p141 and *The Professional Geographer*, vol. 40, no. 4, 1988, p. 470.

Appendix B

Vendors of GKS Software

The following is a list of software vendors who supply GKS implementations. Some of these vendors supply multiple language bindings, device drivers, etc.

ATC
5711 Slauson Avenue, Suite 238
Culver City, CA 90230

Graphic Software Systems Inc.
9590 SW Gemini Drive
Beaverton, OR 97005

Precision Visuals Inc.
6260 Lookout Road
Boulder, CO 80301

SCO Inc.
740C South Pierce Avenue
Louisville, CO 80027-9989

UNIRAS Inc.
50 Mall Road, Suite 212
Burlington, MA 01803

XGKS - Greg Rogers
University of Illinois at Urbana-Champaign
Department of Computer Science
1304 West Springfield Avenue
Urbana, IL 61801

Appendix C

C Program to Read
USGS DEM Data

The following collection of C language functions, making up a program called read_dem, is designed to read the raw data tapes supplied by the USGS containing digital elevation data. These data are discussed in detail in Chapter 5, and some applications are discussed in Chapter 11. The program listing below includes a Makefile (for UNIX and some MS-DOS applications), header files, a main program and several functions. The program should be compiled, linked, and then executed by typing *read_dem*. The program will prompt for the name of the input file. This can be either a raw tape device, such as /dev/rmt12 on UNIX systems, or a file containing the data as extracted from the tape using the appropriate system software (recommended). Microcomputers with nine-track tape drives should be able to use the program, although the software is designed for mini- and mainframe computers with tape drives.

The program will read the tape header and figure out what kind of data are stored, normally either 1:250,000 3 arc-second or 1:24,000 30-meter data. In the latter case, the user will be prompted as to how the map edges will be treated. The choices are to fill the data-less map edges with zeros, and to strip them off by reducing the size of the grid until the edges are free of zeros. This may cause problems when the map edges are oceans. The user also has the option to specify a data subset to extract and write to disk. The output files are dem.dem for the data as elevations in integer meters, nozeros.dem for the edge-stripped array, and dem.doc for the header and adjustment information. Several of the functions, such as convert() and monitor(), are utility functions to adjust for the exact format of the tapes, which are fully documented in the USGS publication cited in Chapter 5. The size of the output files can be rather large.

Function read_dem

```
# Makefile for read_dem : read USGS digital elevation model data

PGM = read_dem

OBJS =   Main.o index.o read_header.o  read_dem.o  conv_buf.o \
         get_value.o  monitor.o  strip_edge.o convert.o

$(PGM) : $(OBJS)
     cc -O $(OBJS) -o $(PGM) -lm

$(OBJS) : Makefile header.h define.h extern.h

/*
/* definition header file for read_dem : generic dem program
 */
#define QUAD7 512
#define QUAD1 512

/*
/* external header file for read_dem : generic dem read program
 */
FILE          *infile, *elev, *docfile;
int            code, usesix, ifedge, echo, image_size, begin_col,
               begin_row, end_col, end_row, this_col, ncols, nrows;
double convert(), get_value(), conv_buf();
double         min, value, yleft, relief;

/*
/* header file for read_dem : generic dem reading program
 */
#include <stdio.h>
#include <math.h>
#include "define.h"
#include "extern.h"

/*
/* Program to read USGS 7 1/2 minute and 1:250,000 digital
/* elevation model tapes, subset them as specified, and to
/* rewrite them as disk files ready for use
/* Author: kcc        Date: 5-88 Last Change: 8-89
 */
#include "header.h"
main()
{
    void            read_header(), read_dem(), strip_edge();
```

```
    /* Read the header from the tape */
    read_header();
    /* Read the dem data */
    read_dem();
    fclose(infile);
    /* Strip the edge from the dem if required */
    if (ifedge) strip_edge();
    printf(" Normal termination: check file dem.doc \n");
}

/*
 * Convert F77 "D" in scientific format to "E"
 */
#include <stdio.h>
FILE            *infile;
double conv_buf()
{
    char            buf[24];
    double          x, atof();
    int             k;
    char            *d, *e;
    d = "D"; e = "E"; fscanf(infile, "%24c", buf);
    for (k = 1; k < 25; k++) if (buf[k] == *d) buf[k] = *e;
    x = atof(buf); return (x);
}

/*
/* Convert arc seconds to degrees,minutes,seconds
 */
#include "header.h"
extern int      code;
extern double   value;
double convert()
{
    int             degrees, minutes, seconds;
    double          floor();
    if (code == 0) {
        /* 1 degree x 1 degree quad */
        degrees = (int) floor(value / 3600.);
            value -= degrees * 3600.;
        minutes = (int) floor(value / 60.);
            seconds = (int) floor(value - (minutes * 60.));
        printf(" %4d degrees %2d minutes %2d seconds\n",
            degrees, minutes, seconds);
        fprintf(docfile, " %4d degrees %2d minutes %2d seconds\n",
            degrees, minutes, seconds);
    } else if (code == 1) {
        /* 7 1/2 x 7 1/2 minute quad - UTM  */
```

```
        printf("%10.1f meters ", value); fprintf(docfile,
            "%10.1f meters ", value);
    } else {
        printf("Error: Value = %10.2f", value); fprintf(docfile,
            "Error: Value = %10.2f", value);
    }
}


/* Get a value in Fortran D format from the file */
#include <stdio.h>
FILE            *infile;
double get_value()
{
    char            buf[20];
    int             k, exp;
    double          x, atof();
    fscanf(infile, "%20c%*2c%2d", buf, &exp);
    x = atof(buf); for (k = 1; k <= exp; k++) x *= 10.0;
    return (x);
}


/*
/* Function to read a standard row header
 */
#include "header.h"
void index()
{
    void monitor();
    int             kt, k, l, skip, pixels, height, z, colnum;
    double          datum, ystart;
    short int       byte;
    char            *zero, c, out;
    zero = "0";
    /* Monitor input until a non-blank is found */
    monitor();
    if (code == 1) {
        fscanf(infile, "%6d", &colnum);
            fscanf(infile, "%6d", &pixels);
        for (k = 1; k < 31; k++) c = getc(infile);
            ystart = conv_buf();
    } else {
        pixels = 1201; for (k = 1; k < 67; k++) c = getc(infile);
    }
    datum = conv_buf();
    for (k = 1; k < 49; k++) c = getc(infile);
    kt = 0;
    if (code == 1) {
        skip = floor((yleft - ystart) / 30.);
```

```
                    if (skip < 0) skip *= -1;
                    for (k = 1; k <= skip; k++) { fprintf(elev, " 0"); kt++; }
            }
        if (this_col < begin_col) {
                for (k = 1; k <= pixels; k++)
                    fscanf(infile, "%6d", &z);
            } else {
                for (k = 1; k <= pixels; k++) { fscanf(infile, "%6d", &z);
                    if ((k <= end_row) && (k >= begin_row)) {
                        height = (int) floor(z + datum);
                        if (usesix) fprintf(elev, "%6d", height);
                        else fprintf(elev, "%5d", height); kt++;
                    }
                }
                kt = image_size - kt;
                if (kt > 0) { for (k = 1; k <= kt; k++)
                    fprintf(elev, " 0"); }
            }

        fprintf(elev, "\n");
        return;
}


/*
 * Search input for start of new record
 */
#include <stdio.h>
FILE            *infile;
void monitor()
{
    char            c, *one;
    one = "1";
    do { c = getc(infile); } while (c != *one);
    return;
}


/*
/* Function to read the dem itself and prompt for options
 */
#include "header.h"
void read_dem()
{
    void index();
    int             kt, kc, this, knt, k, l;
    char            *zero, c;
    zero = "0";
    printf(" Enter 1 if you want the edges removed,
        0 will pad with zeros :");
```

```
        scanf("%d%*1c", &ifedge);
        if (code == 0) {
            printf("\n Enter starting row number for subsample: ");
            scanf("%d", &begin_row);
            printf("\n Enter ending row number for subsample  : ");
            scanf("%d", &end_row);
            printf("\n Enter starting column number for subsample: ");
            scanf("%d", &begin_col);
            printf("\n Enter ending column number for subsample  : ");
            scanf("%d", &end_col);
        } else {
            begin_col = 1; begin_row = 1; end_col = ncols;
                end_row = nrows;
        }
        printf("\n Processing rows %4d to %4d", begin_row, end_row);
        printf("\n and columns %4d to %4d\n", begin_col, end_col);
        elev = fopen("dem.dem", "w"); knt = 0;
        for (this_col = 1; this_col <= end_col; this_col++) {
                if ((this_col) % 10 == 0) printf("\n%5d", this_col);
                    knt++;
                index();
        }
        printf("\n Initial processing complete\n");
        fclose(elev);
}


/*
/* Function to read the header portion of the tape, and to
/* use the information to decide (1) if the tape is 7 1/2
/* minute or 1 degree quad; (2) the minimum and maximum
/* elevations and (3) the number of rows and columns in the
/* matrix.
 */
#include "header.h"
void read_header()
{
    int             k, i, j, zone, eastof, exp;
    char            c, filename[20], buf[20];
    double          max, atof(), fabs(), floor(), yright,
        corner[7];
    printf("\n Enter name of data input file:"); scanf("%s",
        filename);
    if ((infile = fopen(filename, "r")) == NULL) {
        printf("\nError reading input file\n"); exit(0);
    }
    docfile = fopen("dem.doc", "w");
    printf("\n *** Reading header from file %s ***\n", filename);
    fprintf(docfile, "\n *** Header from file %s ***\n", filename);
```

```
for (k = 1; k < 145; k++) { c = getc(infile); putchar(c); }
printf("\n");
fprintf(docfile, "\n");
for (k = 1; k < 13; k++) c = getc(infile);
fscanf(infile, "%6d", &code);
if (code == 1) {
    printf("\n 1 to 24,000 scale 30 meter
        7 1/2 minute DEM");
    fprintf(docfile, "\n 1 to 24,000 scale 30 meter
        7 1/2 minute DEM");
    image_size = QUAD7;
} else if (code == 0) {
    printf("\n 1 to 250,000 scale 3 arc second 1 degree DEM");
    fprintf(docfile, "\n 1 to 250,000 scale 3 arc second
        1 degree DEM");
    image_size = QUAD1;
} else {
    printf("\n Probable error, map code does not match");
    fprintf(docfile, "\n Probable error, map code does
        not match");
}
fscanf(infile, "%6d", &zone);
if (code == 1)
    printf("\n UTM system refers to zone %6d", zone);
fprintf(docfile, "\n UTM system refers to zone %6d", zone);
for (k = 1; k < 379; k++) c = getc(infile);
for (i = 0; i < 8; i++) { corner[i] = get_value(); }
printf("\n Locations of quadrangle corners \n");
fprintf(docfile, "\n Locations of quadrangle corners \n");
printf(docfile, "\n Southwest corner\n");
fprintf(docfile, "\n Southwest corner\n");
for (j = 0; j < 2; j++) { value = corner[j]; convert(); }
printf(" \n Northwest corner\n");
fprintf(docfile, " \n Northwest corner\n");
for (j = 2; j < 4; j++) { value = corner[j]; convert(); }
printf(" \n Northeast corner\n");
fprintf(docfile, " \n Northeast corner\n");
for (j = 4; j < 6; j++) { value = corner[j]; convert(); }
printf("\n Southeast corner\n");
fprintf(docfile, "\n Southeast corner\n");
for (j = 6; j < 8; j++) { value = corner[j]; convert(); }
printf("\n"); fprintf(docfile, "\n");
/* Calculate array limits for 7 1/2 minute quad */
if (code == 1) {
    if (corner[0] < 500000.) { yleft = corner[7];
        yright = corner[3];
    } else { yleft = corner[1]; yright = corner[5]; }
    nrows = floor(fabs((yright - yleft)) / 30.) + 1;
```

```
        } else nrows = 1201;
    min = conv_buf(); max = conv_buf();
    printf("\n Minimum elevation %10.2f meters", min);
    fprintf(docfile, "\n Minimum elevation %10.2f meters", min);
    printf("\n Maximum elevation %10.2f meters", max);
    if (max > 9999) usesix = 1; else usesix = 0;
    fprintf(docfile, "\n Maximum elevation %10.2f meters", max);
    printf("\n Relief        %10.2f meters", relief = max - min);
    fprintf(docfile, "\n Relief        %10.2f meters",
        relief = max - min);
    for (k = 1; k < 73; k++) c = getc(infile);
    fscanf(infile, "%6d", &ncols);
    printf("\n File contains %6d columns\n to be read
        south to north", ncols);
    fprintf(docfile, "\n File contains %6d columns\n to be read
        south to north", ncols);
    printf(" and %6d rows west to east", nrows);
    fprintf(docfile, " and %6d rows west to east", nrows);
    printf("\n\n  *** Header record complete *** \n");
    fprintf(docfile, "\n\n  *** Header record complete *** \n");
    fclose(docfile);
}


/*
/* If the user wants to remove the zero edges, remove them
 */
#include "header.h"
void strip_edge()
{
    int             changed, i, j, begin_col, end_col, begin_row,
                        end_row;
    char            z[QUAD7][QUAD7], *zero, *one;
    FILE            *docfile, *outfile;
    int             zz[QUAD7], found_a_zero, elevation;
    int             blank_row, blank_col;
    zero = "0"; one = "1";
    printf("\n The non-square edges are being removed, output
        is to nozeros.dem\n");
    printf(" check the file dem.doc for the new dem size, etc.\n");
    if ((infile = fopen("dem.dem", "r")) == NULL)
        { printf("\nError reading input file\n"); exit(0); }
    ncols = image_size;
    for (i = 0; i < nrows; i++) {
        for (j = 0; j < ncols; j++) {
            fscanf(infile, "%d", &elevation);
             if (elevation == 0) z[i][j] = *zero;
            else z[i][j] = *one;
        }
```

```
}
begin_col = 0; begin_row = 0; end_col = ncols - 1;
    end_row = nrows - 1;
/* Preprocess to remove entire rows of zeros */
blank_row = 1;
do {
    for (j = begin_col; j <= end_col; j++)
        if (z[begin_row][j] == *one) blank_row = 0;
    if (blank_row) begin_row++;
} while (blank_row); blank_row = 1;
do {
    for (j = begin_col; j <= end_col; j++)
        if (z[end_row][j] == *one) blank_row = 0;
    if (blank_row) end_row--;
} while (blank_row);
/* Preprocess to remove entire cols of zeros */
blank_col = 1;
do {
    for (i = begin_row; i <= end_row; i++)
        if (z[i][begin_col] == *one) blank_col = 0;
    if (blank_col) begin_col++;
} while (blank_col); blank_col = 1;
do {
    for (i = begin_row; i <= end_row; i++)
        if (z[i][end_col] == *one) blank_col = 0;
    if (blank_col) end_col--;
} while (blank_col);
/*
 * Now keep reducing the number of rows and columns until the
 * whole array is data
 */
do { changed = 0; found_a_zero = 0;
    for (i = begin_row; i <= end_row; i++)
        if (z[i][begin_col] == *zero) found_a_zero = 1;
    if (found_a_zero) { begin_col++; changed = 1; }
        found_a_zero = 0;
    for (i = begin_row; i <= end_row; i++)
        if (z[i][end_col] == *zero) found_a_zero = 1;
    if (found_a_zero) { end_col--; changed = 1; }
        found_a_zero = 0;
    for (j = begin_col; j <= end_col; j++)
        if (z[begin_row][j] == *zero) found_a_zero = 1;
    if (found_a_zero) { begin_row++; changed = 1; }
        found_a_zero = 0;
    for (j = begin_col; j <= end_col; j++)
        if (z[end_row][j] == *zero) found_a_zero = 1;
    if (found_a_zero) { end_row--; changed = 1; }
} while (changed);
```

```c
/* Open the file to get the new data */
outfile = fopen("nozeros.dem", "w");
/*
 * Now rewind the file, and write the non-zero part
 */
rewind(infile);
for (i = 0; i < nrows; i++) {
    for (j = 0; j < ncols; j++)
        fscanf(infile, "%d", &zz[j]);
    if ((i >= begin_row) && (i <= end_row)) {
        for (j = begin_col; j <= end_col; j++) {
            if (usesix) fprintf(outfile, "%6d", zz[j]);
            else fprintf(outfile, "%5d", zz[j]);
        }
    }
}
/* Close the output file */
fclose(outfile);
/* Update the .doc file */
docfile = fopen("dem.doc", "a");
fprintf(docfile, "DEM has been stripped of zeros
    at the edge\n");
fprintf(docfile, "Number of columns is now %d\n",
    end_row - begin_row + 1);
fprintf(docfile, "Number of rows    is now %d\n",
    end_col - begin_col + 1);
/*
 * Don't forget, dem rows go south to north, columns go
    east to west
 */
fprintf(docfile, "\n\n Add %d meters to the lower left
    UTM northing\n", 30 * begin_col);
fprintf(docfile, "\n\n Add %d meters to the lower left
    UTM easting \n", 30 * begin_row);
fprintf(docfile, "\n\n Subtract %d meters from the upper
    left UTM northing\n", 30 * (ncols - end_col));
fprintf(docfile, "\n\n Add %d meters to the upper left
    UTM easting \n", 30 * begin_row);
fclose(docfile);
}
```

Index

290